To Maria

Preface

I have for a number of years taught a course in population genetics for students interested in plant and animal breeding. The objective of the course has been to lay a foundation in population genetics for the concepts of quantitative genetics which are introduced in the last third of the course. I have not been able to find an appropriate text for this purpose.

For a quarter of a century, Falconer's *Introduction to Quantitative Genetics* has been the standard, and excellent, text in that subject. For my purposes, however, this text is not sufficiently detailed in the population genetics basis for quantitative theory.

A number of good texts in population genetics are available, of which Li's *First Course in Population Genetics* is didactically the best. But these texts are directed toward the genetics of natural populations, rather than domestic populations, breeding under human control. They also tend to treat quantitative genetics gingerly, if at all.

I have therefore developed the present text from my teaching notes.

The chapters of this book are labeled "Lectures". Each is intended to correspond approximately to the amount of material which can be covered in a 50-minute lecture. Divisions are, of course, dictated by the natural divisions of the subject matter, and the lectures are therefore not of uniform length. Nevertheless, in so far as possible, an attempt has been made to make the average length a lecture's worth.

Prerequisites for the course from which the book sprang are one course in genetics and one in statistical methods. Review lectures on basic genetic and statistical concepts have been included in the book, for review purposes.

Mathematical formulae engender in many people a near catatonic state of fright, making them impervious to the spoken or written word. There is no way to teach population and quantitative genetics without introducing algebraic expressions. As Earl Green points out in the first chapter of his *Genetics and Probability in Animal Breeding Experiments,* mathematical formulae are merely statements about mathematical relations among entities, written in a shorthand form. To emphasize this, formulae are incorporated into the flow of the discussion in the present text, though usually written on a separate line to enhance clarity. An attempt has been made to introduce formulae as gently as possible, and to explain fully what each means. For a biologist, after all, the importance of a formula is not its algebra, but the biological meaning of that algebra.

At the end of most lectures, one or two exercises have been introduced. Complete and detailed solutions for these exercises are given in an Appendix. The exercises are intended primarily to provide examples of the types of data to which the subject matter of the lecture applies, and to give the student the opportunity to manipulate such data. In a few cases, supplementary points about the theory are introduced via the exercises. The student should be encouraged to work out his own answers to the exercises before turning to the printed solution; this will enable him to gauge his progress in understanding the material.

Countless people have contributed to my understanding of population genetics and to the writing of this book. However, I owe a special debt of gratitude to:

Tom Doolittle and Jan Blackburn for the art work; the staffs of the Joan Staats Library at the Jackson Laboratory, Bar Harbor, Maine, and the Albert Mann Library at Cornell University, Ithaca, New York, for their invaluable assistance in finding references; Drs. Duwayne Englert of Southern Illinois University, Carbondale, and Wyman Nyquist, Purdue University, West Lafayette, Indiana, for reading and commenting upon the text; Dr. James F. Crow of the University of Wisconsin, Lansing, for his very detailed commentary; Dr. C. C. Li of the University of Pittsburgh, Pittsburgh, Pennsylvania, for his comments and encouragement; Dr. Dale Van Vleck of Cornell University for multiple acts of assistance: he introduced me to the publisher, sponsored my sabbatical leave at Cornell, during which most of the writing took place, and improved the text with his comments and suggestions; and to my wife, Maria, for her patience and her proof-reading skills.

Any virtues which this book may possess arise from their help; its faults are my responsibility.

DONALD P. DOOLITTLE

Contents

Part I The Hardy-Weinberg Law

Lecture 1 The Hardy-Weinberg Law

The basic law of Mendelian genetics is the law of segregation. It allows us, if we know the genotypes of a pair of parents, to predict the genotypes, in their proportions, that will be found among the offspring of those parents. For example, if both parents are heterozygous *Aa*, we can predict that the three genotypes *AA*, *Aa* and *aa*, will occur among their offspring in an approximate 1: 2:1 ratio.

In a similar manner, the Hardy-Weinberg law is the basic law of population genetics. It allows us to predict the genotypic frequencies in the offspring generation of a population from those of the parent generation. The population is assumed to reproduce by random pairing of genotypes in the parents; Mendelian segregation then controls the frequencies of offspring genotypes produced by any given pair of parental genotypes.

The Hardy-Weinberg law derives its name from the two investigators who independently published it in 1908: George Hardy, a British mathematician, and Wilhelm Weinberg, a German physician. The central principle of the law had actually been stated by the American geneticist, W. E. Castle, as early as 1903 (Li 1967a), but Castle failed to generalize it.

Weinberg's 1908 paper presenting the law was only one of a number of valuable contributions he made to genetics, based on his observations as a practicing obstetrician (Stern 1962). Weinberg's other contributions included a method to estimate the frequency of monozygotic twins in man, and the method of "casting out probands" to correct for errors of ascertainment, as well as further papers on the subject of the Hardy-Weinberg law. But his original contribution to this law was little known, at least to English speaking geneticists, until Stern (1943) published a translation of relevant portions of the 1908 paper.

Hardy's effort was occasioned by his acquaintance with R. C. Punnett, a pioneer British Mendelian geneticist. In those early years of Mendelian genetics, opponents claimed that the law of segregation would require that 1:2:1 genotypic ratios, and 3:1 phenotypic ratios, be characteristic of most traits in any population. Because few traits showed anything like such ratios, the generality of Mendelian inheritance was called into doubt. Punnett challenged Hardy to show that Mendel's laws would not necessarily result in such ratios; and Hardy proceeded to do so.

Let us derive the law, using a somewhat more general argument than Hardy used. Imagine a population of diploid individuals in which a pair of alleles, *A* and *a*, are segregating. Three genotypes, *AA*, *Aa*, and *aa*, are then possible. We assign to these genotypes arbitrary relative frequencies D_t, $2H_t$, and R_t, respectively, in generation t of the population. The only restrictions on these frequencies are that they cannot be less than 0 or greater than 1, and that, since only the three geno-

Table 1.1. Derivation of the Hardy-Weinberg law

Matings	Frequency	Offspring genotypes		
		AA	*Aa*	*aa*
AA × *AA*	D_t^2	D_t^2		
AA × *Aa*	$2D_tH_t$	D_tH_t	D_tH_t	
AA × *aa*	D_tR_t		D_tR_t	
Aa × *AA*	$2D_tH_t$	D_tH_t	D_tH_t	
Aa × *Aa*	$4H_t^2$	H_t^2	$2H_t^2$	H_t^2
Aa × *aa*	$2H_tR_t$		H_tR_t	H_tR_t
aa × *AA*	D_tR_t		D_tR_t	
aa × *Aa*	$2H_tR_t$		H_tR_t	H_tR_t
aa × *aa*	R_t^2			R_t^2
	1	$(D_t+H_t)^2$	$2(D_t+H_t)(H_t+R_t)$	$(H_t+R_t)^2$

types can occur,

$$D_t+2H_t+R_t=1.$$

Now we allow the individuals of generation t to mate at random among themselves. The frequencies of parental genotypic pairs, and of the consequent offspring genotypes produced by each, are summarized in Table 1.1.

The frequency of each parental genotypic combination is simply the product of the frequencies of the genotypes of the two parents. For example, the frequency of *Aa* x *Aa* is $(2H_t)(2H_t)=4H_t^2$. Mendelian segregation ratios, applied to these frequencies, yield the frequencies of offspring genotypes produced. For example, we expect a 1: 2:1 segregation of *AA*: *Aa*: *aa*, from the *Aa* x *Aa* mating, therefore, such matings should contribute H_t^2 *AA*, $2H_t^2$ *Aa* and H_t^2 *aa* to the offspring generation.

The column sums in Table 1.1 represent the overall frequencies of each genotype in generation t+1. A modicum of simple algebra is necessary to simplify these sums. Let us first deal with the sum of the mating frequencies column. This is

$$\begin{aligned}\Sigma\,(\text{freq.}) &= D_t^2+2D_tH_t+D_tR_t+2D_tH_t+4H_t^2+2H_tR_t+D_tR_t+2H_tR_t+R_t^2\\ &= D_t^2+4D_tH_t+2D_tR_t+4H_t^2+4H_tR_t+R_t^2,\end{aligned}$$

which takes the form

$$a^2+2ab+2ac+b^2+2bc+c^2=(a+b+c)^2,$$

with $a=D_t$, $b=2H_t$ and $c=R_t$. Hence,

$$\Sigma\,(\text{freq.})=(D_t+2H_t+R_t)^2=1, \quad (1.1)$$

because $D_t+2H_t+R_t=1$. The total frequency of matings must be 1, regardless of individual values of the genotypic frequencies. Thus, this sum reassures us that we have included all the possible parental genotypic combinations at the correct frequencies.

The frequencies of the three genotypes in generation t+ are the totals of the *AA*, *Aa* and *aa* columns in Table 1.1.

$$D_{t+1} = D_t^2 + 2D_tH_t + H_t^2 = (D_t + H_t)^2;$$
$$2H_{t+1} = 2D_tH_t + 2D_tR_t + 2H_t^2 + 2H_tR_t$$
$$= 2D_t(H_t + R_t) + 2H_t(H_t + R_t)$$
$$= 2(D_t + H_t)(H_t + R_t); \text{ and}$$
$$R_{t+1} = H_t^2 + 2H_tR_t + R_t^2 = (H_t + R_t)^2. \quad (1.2)$$

Now let us consider the frequencies of the alleles *A* and *a* in generation t; we will use p_t and q_t to represent these frequencies, respectively. Because all the genes in *AA* homozygotes, half of those in *Aa* heterozygotes, and none of those in *aa* homozygotes are *A* genes,

$$p_t = 1(D_t) + 0.5(2H_t) + 0(R_t) = D_t + H_t; \quad (1.3a)$$

by similar reasoning,

$$q_t = 0(D_t) + 0.5(2H_t) + 1(R_t) = H_t + R_t. \quad (1.3b)$$

Note that

$$p_t + q_t = D_t + H_t + H_t + R_t = D_t + 2H_t + R_t = 1.$$

Again, p+q must equal 1 if *A* and *a* are the only alleles in the population, providing a check on our calculations.

We now see that the offspring genotypic frequencies are simple functions of the frequencies of the two alleles in the parent generation. The frequencies of offspring genotypes are, from Eqs. (1.2) and (1.3):

$$D_{t+1} = (D_t + H_t)^2 = p_t^2;$$
$$2H_{t+1} = 2(D_t + H_t)(H_t + R_t) = 2p_tq_t;$$
$$R_{t+1} = (H_t + R_t)^2 = q_t^2. \quad (1.4)$$

This implies, of course, that parental genotypic frequencies are irrelevant in determining genotypic frequencies among the offspring. The latter are determined by parental allelic frequencies; any two populations which have different allelic frequencies among the parents will produce different offspring genotypic distributions.

Here, then, was Hardy's answer to the critics of Mendelian inheritance. Mendel's law of segregation does not require all populations to come to a 1: 2:1 genotypic ratio. A distribution is attained which will differ from population to population, and from trait to trait within the same population, depending on allelic frequencies. A footnote to Hardy's paper states that even Udney Yule, one of the most vehement critics of Mendelian genetics, admitted the cogency of this argument.

It also follows, of course, that if two populations have the same allelic frequencies, they will produce offspring in the same genotypic ratio, even if the genotypic distributions of the parents differ. Consider four populations whose genotypic ratios D_t: $2H_t$: R_t are

0.60: 0 : 0.40

0.20: 0.80: 0

0.40: 0.40: 0.20

0.36: 0.48: 0.16.

In all four populations, despite wide disparities in genotypic frequencies in generation t, allele frequencies are $p_t = 0.6$, $q_t = 0.4$. Therefore, all four populations will yield offspring in the Hardy-Weinberg ratio

$$p_t^2 : 2p_tq_t : q_t^2 = 0.36 : 0.48 : 0.16.$$

In any one of these populations,

$$D_{t+1} = 0.36 = p_t^2,\ 2H_{t+1} = 0.48 = 2p_tq_t,\ R_{t+1} = 0.16 = q_t^2;$$

The same ratio will recur in generation t+2.

From Eqs. (1.3),

$$p_{t+1} = p_t^2 + p_tq_t = p_t(p_t + q_t) = p_t;$$
$$q_{t+1} = p_tq_t + q_t^2 = (p_t + q_t)q_t = q_t.$$

In Table 1.2, we have substituted p_t^2 for D_t, $2p_tq_t$ for $2H_t$, and q_t^2 for R_t in Table 1.1. Thus, Table 1.2 represents the parental genotypic frequencies in generation t+1, and the offspring frequencies in t+2, expressed as functions of generation t allele frequencies. From this table, we see that the offspring ratio is, indeed, the Hardy-Weinberg ratio, even when the parents are already in that ratio. For example,

$$D_{t+2} = p_t^4 + 2p_t^3q_t + p_t^2q_t^2 = p_t^2(p_t^2 + 2p_tq_t + q_t^2) = p_t^2,$$

Table 1.2. Hardy-Weinberg equilibrium

Matings	Frequency	Offspring genotypes		
		AA	*Aa*	*aa*
AA × *AA*	p_t^4	p_t^4		
AA × *Aa*	$2p_t^3q_t$	$p_t^3q_t$	$p_t^3q_t$	
AA × *aa*	$p_t^2q_t^2$		$p_t^2q_t^2$	
Aa × *AA*	$2p_t^3q_t$	$p_t^3q_t$	$p_t^3q_t$	
Aa × *Aa*	$4p_t^2q_t^2$	$p_t^2q_t^2$	$2p_t^2q_t^2$	$p_t^2q_t^2$
Aa × *aa*	$2p_tq_t^3$		$p_tq_t^3$	$p_tq_t^3$
aa × *AA*	$p_t^2q_t^2$		$p_t^2q_t^2$	
aa × *Aa*	$2p_tq_t^3$		$p_tq_t^3$	$p_tq_t^3$
aa × *aa*	q_t^4			q_t^4
	1	p_t^2	$2p_tq_t$	q_t^2

and so forth. A population in the Hardy-Weinberg genotypic ratio in one generation will produce offspring in the same ratio in the next.

Any population will reach the Hardy-Weinberg ratio in generation t+1, and remain in that ratio in generations t+2, t+3, etc. All of the populations above will produce the ratio 0.36: 0.48: 0.16 from generation t+1 on, as long as they continue to reproduce under Hardy-Weinberg conditions. This has been termed Hardy-Weinberg equilibrium (Wentworth and Remick, 1916).

Allele frequencies among the zygotes of generation t+1 will be the same as among their generation t parents:

$$p_{t+1(z)} = p_t^2 + 0.5(2p_tq_t) = p_t^2 + p_tq_t = p_t(p_t + q_t) = p_t;$$

$$q_{t+1(z)} = 0.5(2p_tq_t) + q_t^2 = q_t(p_t + q_t) = q_t; \qquad (1.5)$$

where $p_{t+1(z)}$ is the *A* allele frequency among the zygotes of generation t+1. If allele frequencies are constant throughout the life of generation t+1, when that generation mates at random, it will again produce zygotes in the same Hardy-Weinberg ratio, p_t^2: $2p_tq_t$: q_t^2. The population will be in Hardy-Weinberg equilibrium.

Suppose, however, that allele frequencies change during generation t+1, so that adults of that generation have *A* alleles in the frequency

$$p_{t+1(m)} \neq p_{t+1(z)} = p_t.$$

Random mating among these adults will produce generation t+2 zygotes in the ratio

$$p_{t+1(m)}^2 : 2p_{t+1(m)}q_{t+1(m)} : q_{t+1(m)}^2.$$

This is a Hardy-Weinberg ratio, but one based on $p_{t+1(m)}$ rather than on p_t.

Because genotypic frequencies change, there is no equilibrium. Genotypic frequencies can remain the same from generation to generation only if allele frequencies remain the same. Constant allele frequency is an essential condition for Hardy-Weinberg equilibrium.

However, each generation produces zygotes in a Hardy-Weinberg ratio based on the allele frequencies of their parents. This ratio is solely a consequence of random mating. If parents mate at random, offspring zygotes will occur in the Hardy-Weinberg ratio, even if the genotypic ratio may change at later stages of the offspring generation. Random mating is an essential condition for the Hardy-Weinberg ratio; constant allele frequency is not.

In deriving the Hardy-Weinberg law, we tacitly assumed constant allele frequencies, simply by assuming that none of the forces which cause allele frequency to change were operative. This is only one of a set of rather narrowly limited assumptions implicit in the derivation. Hardy himself admitted the limitations, as did Weinberg; and the latter, in his 1909 and 1910 papers, made some attempts to broaden the limits. In the next lecture, we will make these assumptions explicit.

Lecture 1 Exercises

1. The four populations

a) 0.60: 0 : 0.40

b) 0.20: 0.80: 0

c) 0.40: 0.40: 0.20

d) 0.36: 0.48: 0.16

are stated in the text to produce the same genotypic distribution in generation $t+1$. Prepare tables in the form of Table 1.1 to show that this is true.

The ratios D_t: $2H_t$: R_t are given below for five populations. Determine p_t and q_t and the equilibrium genotypic proportions for each population.

a) 0.50: 0 : 0.50

b) 0.50: 0.50: 0

c) 0.09: 0.10: 0.81

d) 0.36: 0.16: 0.48

e) 0.45: 0.45: 0.10

(N.B.: In each case, $p_t+q_t=p_t^2+2p_tq_t+q_t^2=1$. Use these relationships to check your results.)

3. Much of the algebra involved in population genetics consists of manipulations based upon the fact that $p+q=1$ (and that therefore, $p=1-q$, $q=1-p$). Use this fact to verify that

a) $p+2q=1+q=2-p$

b) $p^2-q^2=p-q$

c) $p^2+q^2=1-2pq$

d) $(p-q)^2=1-4pq$

e) $p^2+4pq+4q^2=(1+q)^2$

Lecture 2 Conditions for Hardy-Weinberg Equilibrium

We have derived the Hardy-Weinberg law under a set of assumed conditions, only a few of which were stated explicitly. We must now make all of these assumptions explicit, defining terms clearly.

We introduced, without defining, the concept of a population of organisms. The population geneticist defines a population as a set of freely interbreeding organisms. This definition implies that all members of the population are of the same species; otherwise, free interbreeding would not be possible. For a population in nature, it also implies geographic propinquity of members of the population; individuals living too far apart cannot freely interbreed. "Too far apart", of course, will be conditioned by the nature of the species; it may be a much shorter distance for a sessile plant than for a mobile animal, for example. Populations living under domestication can be limited in breeding range, in the most general meaning of that term. A plant or animal breeder may restrict the breeding range of a given individual to other members of the same breed or variety, even though the individual could breed fertilely with any other member of the same species. At the same time, domesticated plants and animals can, with man's help, transcend normal geographic limitations on breeding range. Individuals, or their gametes, under artificial insemination, can be transported across or even between continents to breed.

We assumed, in deriving the law, that breeding occurred only within the population, that is, that all of the parents of one generation of the population were members of an earlier generation of the same population. In general, we will continue to assume breeding within the population. When it becomes necessary to transgress this limitation, as in discussing crossing or migration, we will clearly indicate that we are doing so.

We also assumed that generations were discrete, that is, that members of generation t mated only with other members of generation t. The model is that of an annual plant, with this year's generation dead and gone before next year's can come to flower. Many populations are not subject to this kind of restriction. An individual from generation t may still be alive and fertile to mate with members of generation $t+1$, $t+2$, etc. Overlapping generations are probably more common than discrete generations in nature. In domestic populations, a superior individual will often be retained to breed with descendant generations.

Nonetheless, the assumption of discrete generations will greatly simplify the discussion of many of the factors whose effects we will be investigating, without any great sacrifice in generality (Crow and Kimura 1970, Hill 1972c, 1979). We will therefore make this assumption in most of our discussions.

A given generation number applies to the zygotes of that generation, the adults which develop from those zygotes, and the gametes those adults produce.

That is to say, generation t zygotes develop into generation t adults, which produce generation t gametes. Generation t+1 starts with the zygotes formed by the union of generation t gametes.

We assumed Mendelian segregation in deriving the Hardy-Weinberg law. The population was regarded as a congeries of matings, the proportions of genotypic combinations being determined by the parental genotypic ratio. The genotype of the offspring produced by each mating was then determined by Mendelian segregation from the parental genotypes. We will continue to make this assumption in the discussions which follow. Segregation is modified in the case of sex-linked, partially sex-linked and holandric genes, but the basic mechanism is still Mendelian. Extrachromosomal genes, if such exist, may not segregate according to Mendel's law; some types of aneuploidy may also cause non-Mendelian segregation. Such conditions will not fall within the purview of our discussions.

In deriving the law, we allowed "the individuals of generation t to mate at random," without defining what "to mate at random" meant. The model of random mating which we assumed can be described as follows.

We choose an individual from generation t of the population "at random", which is to say that our choice is dictated purely by chance; every member of generation t has equal probability of being chosen. The probability that we will choose an individual of any given genotype, say *AA*, is therefore the frequency of that genotype in generation t, D_t. We then choose a second individual, again at random, to mate with this first. The first is assumed to be still available for the second choice, so that selfing is possible. What is more important, the choice of the first individual does not change the genotypic frequencies; the probability of choosing *AA* on the second choice is still D_t. The choices are independent, and the probability of choosing any pair of genotypes is the product of the frequencies of the two genotypes. From this pair of mates we produce a single offspring, whose genotypic probabilities are dictated by Mendelian segregation from the parental genotypes. Innumerable repetitions of this procedure produce generation t+1 of the population.

Random mating as described above implies randomization in two phases. Mates are chosen and paired at random; the random choice of a gamete from each mate is implicit in Mendelian segregation. We can combine these in a single randomization phase. Suppose we obtain equal numbers of gametes from each member of generation t, and pool them. Assuming equal fertility, each individual will contribute an equal number of gametes to the pool. The pool will then contain *A* and *a* gametes in the ratio p_t: q_t. Gamete frequencies are equal to the frequency of the corresponding allele, because each gamete carries one gene.

If we now draw a random gamete from the pool, the probability of choosing a gamete carrying a particular allele will be the frequency of that gamete. If we unite pairs of randomly drawn gametes, the probability of producing a given genotype will be the product of the frequencies of the gametes in that genotype. Innumerable repetitions of this procedure will give the results summarized in the modified Punnett square of Table 2.1. The genotypic ratio will be the Hardy-Weinberg ratio, p_t^2: $2p_tq_t$: q_t^2, the same ratio we derived from random mating. Random union of gametes is equivalent to random mating of individuals.

Table 2.1. Random union of gametes

First gamete	Second gamete	
	A	a
	p_t	q_t
A p_t	$p_t^2 AA$	$p_t q_t Aa$
a q_t	$p_t q_t aA$	$q_t^2 aa$

As we consider the effects of changing the assumptions under which we derived the Hardy-Weinberg law, the equivalence of random mating and random union of gametes will provide us with a convenient shortcut. In many cases, we will not have to consider the effect of changes on genotypic ratios at all, because, given random mating, there is a 1:1 relationship between the gamete array and the ratio of genotypes.

Under certain circumstances, e.g., self sterility, natural populations will not mate at random. A breeder may also impose nonrandom mating on a domesticated population. We will investigate the consequences of such nonrandom mating.

Random union of gametes implies that any two gametes from any two individuals can successfully unite to form an offspring. We know, however, that in most species, two types (sexes) of gamete occur, and that only the union of gametes of opposite sex will form a viable zygote. In most animal species, the two sexes exist in separate individuals. Even in plants, where both often come from the same individual, male and female gametes are formed. We must, then, explore the effects of separation of the sexes upon equilibrium.

When sexes are separate, there may be an hereditary mechanism for determining sex, such as the sex chromosomes of many higher animal species. Do genes carried on the sex chromosomes reach equilibrium in the same manner as autosomal genes?

We derived the Hardy-Weinberg law for two alleles in a diploid. What are the effects of multiple alleles, of polyploidy? We also considered only one locus; how do two or more loci approach equilibrium simultaneously?

We have seen that it was necessary to assume constant allele frequencies to produce Hardy-Weinberg equilibrium. This meant assuming equal viability, fertility and fecundity of all genotypes, and ruling out mutation and migration. What effect do inequalities of viability and reproductive capability have? Do populations subject to mutation and migration reach equilibrium?

Finally, we implicitly assumed, by equating probabilities with frequencies, that our population was infinitely large. This assumption was also necessary to avoid changes in allele frequency due to sampling error. Because no real population can be infinitely large, we must examine the effects of finite size.

The above assumptions indicate that the Hardy-Weinberg equilibrium can apply to a population only under rather narrowly defined conditions. Before examining the effects of changing these conditions, we must first ask whether any real population can exist to which the Hardy-Weinberg assumptions do apply.

Lecture 2 Exercises

1. The student can simulate random mating, using any of a number of randomization devices (dice, poker chips, tables of random numbers, computer programs for generating random results). The device should be arranged to provide three classes ("genotypes") of results in an arbitrarily predetermined ratio D_t: $2H_t$: R_t. A series of trials are made, the genotype from each being recorded. Pairs of successive genotypes represent matings. Each student should make at least a hundred matings to obtain a fair sample. If all students in the class use the same ratio of genotypes, pooled results from the whole class should give an even better representation of a large population.

2. Random union of gametes can be simulated using two result classes in the ratio p_t: q_t.

Lecture 3 Applications of the Hardy-Weinberg Law

Can any real population satisfy the stringent conditions outlined in Lecture 2, under which our theoretical population reached Hardy- Weinberg equilibrium? Probably not; but many real populations have been shown to produce genotypes, with respect to some traits at least, in a ratio not significantly different from p^2: 2pq: q^2.

J.F. Crow (1985, personal communication) emphatically points out that the existence of this ratio does not prove that a population is in Hardy-Weinberg equilibrium. If allele frequencies are p and q among the gametes, and if gametes unite at random, zygote genotypes will occur in the Hardy-Weinberg ratio, even if the population is not in equilibrium.

This is undoubtedly true; yet there is some justification for testing for a Hardy-Weinberg ratio. Deviations from this ratio show clearly that the population is not in Hardy-Weinberg equilibrium. Forces preventing equilibrium by changing allele frequencies often also cause changes in the genotypic ratio (for example, migration, or selection due to differential viability), especially if the ratio is measured when the population is near reproductive age. Departures from the Hardy-Weinberg ratio may indicate the nature of the forces causing the deviation. For example, an excess of homozygotes and a dearth of heterozygotes may indicate inbreeding. On the other hand, the existence of a Hardy-Weinberg ratio at least indicates that mating is random. Thus, we can regard testing for Hardy-Weinberg ratios as at least a partial test for Hardy-Weinberg equilibrium.

It would be fatuous to seek Hardy-Weinberg equilibrium in a population or for a trait which we know does not fulfill the required conditions. We need a diploid population of reasonably large size, not known to be subject to migration or to nonrandom mating. The trait must be inherited by two alleles at a single autosomal locus, nearly neutral to selection, and showing no evidence of excessive mutation rates. For such a trait in such a population, observed numbers of the different phenotypes can be compared to numbers expected on the basis of allele frequencies, allowance being made for sampling error.

It is easier to make these comparisons using a trait where there is codominance or partial dominance, so that the three genotypes, *AA*, *Aa*, and *aa*, can be distinguished by phenotype. Suppose, then, that we have a population of diploid organisms in which a pair of alleles, *A* and *a*, are segregating at an autosomal locus, with the three genotypes differing in phenotype. We draw a random sample of N_{Tt} individuals from generation t of this population, and count N_{Dt} *AA*, N_{2Ht} *Aa* and N_{Rt} *aa* individuals in this sample. Of course, all members of the population must have one or another of these three genotypes; therefore,

$$N_{Dt} + N_{2Ht} + N_{Rt} = N_{Tt}.$$

And because our sample was drawn at random,

$$N_{Dt}/N_{Tt} = \hat{D}_t, \text{ etc.;}$$

the sample frequencies of the three genotypes are valid estimates of the frequencies in the population.

We can regard the population that actually exists as a random sample from a theoretical infinite population of all individuals that could have existed. The observed genotypic frequencies are estimates of the frequencies in the infinite population, whether we examine all members of the existing population, or only a subsample from it. As long as the subsample and the sample are both drawn at random, statistics from the subsample estimate the population parameters.

From our sample, then, we can estimate the population frequencies of *A* and *a* alleles. Since each *AA* carries two *A* genes, and each *Aa* one, we have $(2N_{Dt}+N_{2Ht})$ *A* genes in the sample, among a total of $2N_{Tt}$ genes. Then

$$\hat{p}_t = (2N_{Dt}+N_{2Ht})/2N_{Tt};$$
$$\hat{q}_t = (N_{2Ht}+2N_{Rt})/2N_{Tt}. \qquad (3.1)$$

These estimates do not depend on any assumptions about the genotypic distribution. Equations (3.1) give the sample allele frequencies, and thus estimate the allele frequencies in the population, whether or not genotypes are in the Hardy-Weinberg ratio.

These estimates, being based on a finite sample, are subject to sampling variation. The sampling variances for $\hat{p}_t$ and $\hat{q}_t$ are identical, because if $\hat{p}_t$ is greater than p_t, $\hat{q}_t$ must be less than q_t by an equal amount, and vice versa. The variance, derived from the binomial frequency distribution, is

$$\mathrm{Var}(\hat{p}_t) = \mathrm{Var}(\hat{q}_t) = p_t q_t/2N_{Tt}. \qquad (3.2)$$

This variance is a function of the population parameters, p_t and q_t. Therefore, we can never exactly evaluate $\mathrm{Var}(\hat{p}_t)$; we can only estimate it by substituting sample estimates for the parameters in Eq. (3.2).

If the population were in Hardy-Weinberg equilibrium, genotypic frequencies would be in the Hardy-Weinberg ratio, p_t^2: $2p_tq_t$: q_t^2, in generation t. We can estimate the numbers of individuals of each genotype which we would expect to occur from our estimates of allele frequencies:

$$E_{Dt} = \hat{p}_t^2 N_{Tt};$$
$$E_{2Ht} = 2\hat{p}_t\hat{q}_t N_{Tt};$$
$$E_{Rt} = \hat{q}_t^2 N_{Tt}. \qquad (3.3)$$

We can now test for a Hardy-Weinberg ratio in the population by comparing the expected numbers from Eqs. (3.3) to the numbers actually observed.

A simple comparison of the two sets of numbers may be sufficient to convince us that the observed numbers are or are not in the Hardy-Weinberg ratio. However, we will usually want a statistical test of the significance of differences between observed and expected numbers, which will make due allowance for sampling errors. This is provided by the χ^2 ratio,

$$\chi^2 = (N_{Dt} - E_{Dt})^2/E_{Dt} + (N_{2Ht} - E_{2Ht})^2/E_{2Ht} + (N_{Rt} - E_{Rt})^2/E_{Rt}. \tag{3.4}$$

Because three comparisons are being made within the χ^2 test, and we use one degree of freedom to make the sum of the expected numbers equal to N_t, and one to estimate p (or q), the resultant test has one degree of freedom.

The 5% probability value for the χ^2 distribution with d.f. is 3.841. If χ^2 calculated from Eq. (3.4) is greater than 3.841, the probability that the difference between the observed and expected numbers arose by sampling error is less than 1 in 20. The null hypothesis that the observed genotypes were in the ratio p_t^2: $2p_tq_t$: q_t^2 would be rejected, and we would conclude that the population is not in Hardy-Weinberg equilibrium. If, on the other hand, the calculated χ^2 is less than 3.841, the difference between expected and observed numbers of the various genotypes may very well be simply the result of sampling variation. We do not have any reason to reject the null hypothesis but can conclude that, for practical purposes, the population produces the Hardy-Weinberg ratio of genotypes with regard to this trait.

As an example of the application of this test, consider results obtained by Fabricius-Hansen (1940) in a study of MN blood groups among West Greenland Eskimos. The MN blood groups in man show codominant inheritance; group M consists of the $I^M I^M$ homozygotes, group N the $I^N I^N$ homozygotes, while $I^M I^N$ heterozygotes constitute the MN group (Thompson and Thompson 1980). Calculations based on these data are shown in Table 3.1. It is possible to check the cal-

Table 3.1. Testing for Hardy-Weinberg genotypic proportions; codominance. (From Fabricius-Hansen, 1940)

Phenotype	Symbol	Number		
		Observed	Expected	O–E
M	N_{Dt}	485	488.7	−3.7
MN	N_{2Ht}	227	219.6	+7.4
N	N_{Rt}	21	24.7	−3.7
Total	N_{Tt}	733	733.0	

Allele frequency estimates:

$\hat{p}_t = (2N_{Dt} + N_{2Ht})/2N_{Tt} = (970 + 227)/1466 = 0.8165$

$\hat{q}_t = (227 + 42)/1466 = 0.1835$

Estimated variance:

$\mathrm{Var}(\hat{q}_t) = \hat{p}_t\hat{q}_t/2N_{Tt} = (0.8165)(0.1835)/1466 = 0.0001$

Expected numbers:

$E_{Dt} = \hat{p}_t^2 N_{Tt} = (0.8165)^2(733) = 488.7$

$E_{2Ht} = 2(0.8165)(0.1835)(733) = 219.6$

$E_{Rt} = (0.1835)^2(733) = 24.7$

Test of significance:

$\chi^2 = (-3.7)^2/488.7 + (7.4)^2/219.6 + (-3.7)^2/24.7 = 0.8316$

culations of allele and genotype frequencies to avoid arithmetic errors. The allele frequency estimates $\hat{p}_t$ and $\hat{q}_t$ should sum to one. The sum of the expected numbers of genotypes, E_{Dt}, E_{2Ht} and E_{Rt}, should be N_{Tt}. And the differences between observed and expected numbers in Eq. (3.4) should sum to zero.

The observed and expected numbers differ so little, in Fabricius-Hansen's data, that the χ^2 test is hardly necessary. Nonetheless, it is calculated in Table 3.1 for the sake of completeness. The calculated χ^2 value is 0.8582, much less than the critical ($p=0.05$) value of 3.841. In fact, there is approximately a 40% probability that a χ^2 as large as or larger than 0.8582 will arise by sampling variation. We can therefore conclude that, with regard to the MN blood group genotypes, the West Greenland population sampled by Fabricius-Hansen produces offspring in the Hardy-Weinberg ratio.

It should be noted that the χ^2 test is applied to expected and observed *numbers* of the three genotypes. χ^2 is not valid for the comparison of frequencies.

The χ^2 test cannot be directly applied if the number of individuals of one or more genotypes is very small. Often, in natural populations, one allele may be very rare, so that few individuals homozygous for that allele will be expected. Under these conditions, the test procedures described above cannot be employed. Li (1976) reviews a number of alternative methods which can be used in such cases.

The above procedures apply when there is codominance or partial dominance, so that the three genotypes can be distinguished by phenotype. Suppose, however, that the *A* allele is completely dominant to *a*. There are then only two phenotypes, Dominants and Recessives. All Recessives will be *aa* in genotype, but a Dominant may be either *AA* or *Aa*; we cannot distinguish between the latter genotypes by their phenotypes.

The number of Recessives in generation t is N_{Rt}. But the number of Dominants is $N_{Dt}+N_{2Ht}$. Because we cannot separate these two genotypes, we cannot use Eqs. (3.1) to estimate allele frequencies. We can only estimate p_t and q_t by assuming that genotypes are in Hardy-Weinberg proportions. Making this assumption,

$$R_t = q_t^2,$$

and we can estimate (Bernstein 1925)

$$\hat{q}_t' = (N_{Rt}/N_{Tt})^{0.5},$$
$$\hat{p}_t' = 1 - \hat{q}_t'. \tag{3.5}$$

These allele frequency estimates have a greater variance than the estimates derived from Eqs. (3.1). The variance of $\hat{q}_t'$ is

$$\begin{aligned} \mathrm{Var}(\hat{q}_t') &= (1-q_t^2)/4N_{Tt} = (p_t^2+2p_tq_t)/4N_{Tt} \\ &= p_t^2/4N_{Tt} + 2p_tq_t/4N_{Tt} = p_tq_t/2N_{Tt} + p_t^2/4N_{Tt} \\ &= \mathrm{Var}(\hat{q}_t) + p_t^2/4N_{Tt}. \end{aligned} \tag{3.6}$$

Because neither p_t nor N_{Tt} can be negative, $p_t^2/4N_{Tt}$ must be positive, and $\mathrm{Var}(\hat{q}_t') > \mathrm{Var}(\hat{q}_t)$.

Table 3.2. Testing for Hardy Weinberg genotypic proportions; complete dominance

Phenotype	Number expected	Number observed
M+	$N_{Dt}+N_{2Ht}$	712
M−	N_{Rt}	21
Total	N_{Tt}	733

Allele frequency estimates:

$\hat{q}_t'=(N_{Rt}/N_{Tt})^{0.5}=0.1693$

$\hat{p}_t'=1-\hat{q}_t'=0.8307$

Estimated variance:

$\mathrm{Var}\ (\hat{q}_t')=(1-\hat{q}_t^2)/4N_t=0.0003$

Expected numbers:

$E_{Rt}=(\hat{q}_t')^2N_{Tt}=21.0096$

$E_{Dt}+E_{2Ht}=N_{Tt}-E_{Rt}=711.9904$

There is also a more important disadvantage to estimating q_t from Eq. (3.5), rather than Eq. (3.1). We have had to assume the Hardy-Weinberg genotypic ratio to estimate q_t; therefore, we cannot use the same data to test for this ratio. Aside from arithmetic and rounding errors, the expected number of Dominants and Recessives derived from applying functions of $\hat{q}_t'$ to N_{Tt} will exactly equal the observed numbers of the various genotypes, because the procedures for deriving the expected numbers simply reverse the estimation procedures.

If Fabricius-Hansen had had only anti-M serum, no anti-N, available for testing his Eskimos, both group M and group MN persons would have reacted positively to the anti-M serum. Only group N persons would not, simulating a situation in which the I^M allele was dominant to I^N. The calculation of allele frequency estimates from Eq. (3.5) is shown in Table 3.2 for these data, along with the estimated variance of $\hat{q}_t'$. These can be compared with results from Table 3.1. The estimate of q_t is only about 8% less using Eq. (3.5) than that from Eq. (3.1); but the estimated variance for $\hat{q}_t'$ is about three times as large as that for $\hat{q}_t$.

Expected numbers of Dominants and Recessives are calculated in Table 3.2 for the complete dominance case, to show that they are virtually identical with the observed numbers. No χ^2 test can be calculated in Table 3.2, because such a test would take the form

$$\chi^2=(N_{Dt}+N_{2Ht}-E_{Dt}-E_{2Ht})^2/(E_{Dt}+E_{2Ht}) + (N_{Rt}-E_{Rt})^2/E_{Rt}.$$

There are two comparisons. One degree of freedom is used to estimate allele frequencies, and one to make the total of the expected numbers equal N_{Tt}. None is left to make the test.

In order to test for the Hardy-Weinberg ratio with complete dominance, we need one sample in which to estimate allele frequency, and a second in which to compare expected and observed numbers. Both samples must come from the same population; for example, we might estimate q in generation t and test for the genotypic ratio in generation t+1. An example of this type of test is used as one of the exercises associated with this lecture.

Lecture 3 Exercises

1. In data quoted by Li (1976, pp. 17-18), from a sample of 1029 persons tested, 342 were group M, 500 group MN and 187 group N. Test these data for the Hardy-Weinberg ratio.

2. Suppose that, in the example of Exercise 1, it had not been possible to test for the M antigen, only for the presence or absence of antigen N. What would the best estimate of q, the frequency of I^M, have been? Estimate the variances of $\hat{p}_t$ from Exercise 1, and of $\hat{p}_t'$ from this exercise, and compare the two.

3. Suppose that, in Exercise 2, we have data on the offspring population, again tested only with anti-N serum. In this test, we find among 101 individuals, 67 N+ and 340 N−. Test for the Hardy-Weinberg ratio in this population.

4. In the presence of pink-eyed dilution (*pp*), the genotypes $c^{ch}c^{ch}$, $c^{ch}c$ and *cc* are distinguishable in mice. In a stock homozygous *pp*, we find the numbers of these three genotypes are 457, 444 and 231, respectively. Test for the Hardy-Weinberg ratio.

Lecture 4 Snyder's Ratios

In 1931, Dr. Laurence H. Snyder, a student of human inheritance, became interested in a report (Fox 1931) that people differed in their ability to taste the chemical phenylthiocarbamide (PTC). He tested members of 100 families (Snyder 1931), with results that strongly suggested that PTC taste blindness was inherited as a Mendelian recessive trait. All children from marriages between two nontasters were nontasters, while the frequency of nontaster children from marriages where one parent was a nontaster was greater than from marriages where both parents could taste PTC. Blakeslee and Salmon (1931) reported similar results in another series of 47 families.

Snyder (1932) expanded his sample to 800 families with a total of 2043 children (Table 4.1). Only 5 among 223 children from nontaster x nontaster marriages could taste the chemical, and these could be explained as due to adoption, illegitimacy, new mutations or errors in reporting the ability to taste. This appeared to provide clear proof that taste blindness was due to homozygosity for a recessive gene, *t*. *TT* homozygotes and *Tt* heterozygotes should both be able to taste the chemical.

Had all taster parents been *Tt* heterozygotes, taster x taster marriages should have produced 25%, and taster x nontaster marriages 50%, nontaster children. Snyder observed only 12% and 37% nontaster children, respectively, from these two types of marriage, and realized that this was because some taster parents were *TT* homozygotes. Any marriage in which one or both parents were *TT* could not produce any *tt* children, and their families would dilute the frequency of nontaster children from marriages where the taster parent or parents were *Tt*.

He reasoned, however, that if the genotypic distribution among the parents were in the Hardy-Weinberg ratio, it should be possible to estimate the frequencies of $Tt \times Tt$ and $Tt \times tt$ marriages and so to predict the expected frequencies of nontaster children. His reasoning is summarized in Table 4.2.

Table 4.1. Data illustrating Snyder's ratios (Snyder, 1932)

Marriages		Children		
Type	No.	Taster	Nontaster	Total
Taster × taster	425	929	130	1059
Taster × nontaster	289	483	278	761
Nontaster × nontaster	86	5	218	223
	800	1417	626	2043

Table 4.2. Snyder's ratios

Marriages		Frequency of *tt* children
Type	Frequency	
Taster × taster:		
TT × *TT*	p^4	0
TT × *Tt*	$4p^3q$	0
Tt × *Tt*	$4p^2q^2$	p^2q^2
	$p^2(1+q)^2$	p^2q^2
Taster × nontaster:		
TT × *tt*	$2p^2q^2$	0
Tt × *tt*	$4pq^3$	$2pq^3$
	$2pq^2(1+q)$	$2pq^3$

Letting p be the frequency of *T* and q of *t*, and assuming a Hardy-Weinberg ratio of genotypes among the parents, the overall frequency of tasters should be $p^2+2pq=1-q^2$, the sum of the frequencies of *TT* and *Tt*. If marriages occurred at random with respect to this trait, the frequency of taster x taster marriages should be $(1-q^2)^2$. Genotypes of taster x taster marriages should be

TT × *TT* with frequency p^4;

TT × *Tt* with frequency $4p^3q$; and

Tt × *Tt* with frequency $4p^2q^2$.

(The frequency of *TT* × *Tt* marriages is the sum of the frequencies, in Table 1.1, of *AA* × *Aa* and *Aa* × *AA* matings. Because *TT* × *Tt* and *Tt* × *TT* marriages would produce the same genotypes in the same proportions among their offspring, we combine them.)

The frequencies of these three types of marriage do indeed sum to $(1-q^2)^2$:

$$p^4+4p^3q+4p^2q^2=(p^2+2pq)^2=(1-q^2)^2.$$

Because $(1-q^2)=(1-q)(1+q)$, we can write this in the form

$$(1-q^2)^2=(1-q)^2(1+q)^2=p^2(1+q)^2.$$

TT × *TT* and *TT* × *Tt* marriages would produce only taster children. Only *Tt* × *Tt* among taster × taster marriages would produce 25% nontaster offspring. The production of *tt* children would be

$$4p^2q^2/4=p^2q^2,$$

and Snyder would expect

$$S_2=p^2q^2/p^2(1+q)^2=q^2/(1+q)^2 \quad (4.1)$$

to be the frequency of nontasters among children from taster × taster marriages.

Similarly, taster × nontaster marriages should be

TT × *tt* with frequency $2p^2q^2$, and

Tt × *tt* with frequency $4pq^3$.

Again, frequencies for matings involving the same two genotypes in reverse order have been combined. The overall frequency of taster × nontaster marriages is then

$$2p^2q^2 + 4pq^3 = 2q^2(p^2 + 2pq) = 2q^2(1 - q^2)$$
$$= 2q^2(1 - q)(1 + q) = 2pq^2(1 + q).$$

TT × *tt* marriages produce no nontaster children, but half the children from *Tt* × *tt* are nontasters. Thus,

$$S_1 = 2pq^3/2pq^2(1 + q) = q/(1 + q) \tag{4.2}$$

should be the frequency of nontaster children from taster × nontaster marriages.

S_1 and S_2 are called "Snyder's ratios". The subscript provides a convenient reminder of the type of marriage to which each ratio applies: S_2 when both parents are dominants, S_1 when one parent is dominant and one recessive. Multiplying each ratio by the total number of children from the appropriate type of marriage gives the expected number of nontaster children from that type of marriage. This can then be compared to the observed number by χ^2 test.

Fisher (1934) correctly observed that this test is not strictly applicable unless all families have exactly the same number of children. Smith (1956) described a modification of the expected number calculation to correct for unequal family sizes. Unless, however, the majority of children observed come from a small number of very large families, the distortion of the χ^2 ratio is not likely to be of importance.

We can use Snyder's 1932 data (Table 4.1) to illustrate the application of the ratios. To estimate S_1 and S_2, we must first estimate q. Since we are dealing with a trait in which there is dominance, we must use $\hat{q}'$ [Eq (3.5)] for this estimate. Snyder estimated q from the frequency of nontasters across both parents and children. We have omitted the generation index t in our description of Snyder's rationale above, partly to avoid confusion with the allele symbol *t*, but also because Snyder's estimate of allele frequencies came from both generation t (the parents) and generation t+1 (the children). Cepellini et al. (1955) have argued that, since the genes in the offspring are only copies of the parental genes, only the latter should be counted in estimating allele frequencies. Parent and offspring genotypes are, of course, correlated, and the sampling error of this estimate from both generations is probably larger than the total number of parents plus children would indicate. But this sampling error must still be less than that for an estimate from either generation alone.

We can estimate q from combined data only if q is the same in both generations. Under Hardy-Weinberg equilibrium it should be, of course; but it may be advisable to compare the overall frequencies of Recessives in the two generations. This can be done by multiplying the parent frequency, N_{Rt}/N_{Tt}, by the total number of children, N_{Tt+1}, to obtain an expected number of Recessive children, E_{Rt+1}. In Snyder's data,

$$N_{Rt}/N_{Tt} = [2(86) + 289]/2(800) = 0.2881.$$
$$E_{Rt+1} = (0.2881)(2043) = 589.$$

N_{Rt+1} is 626. The χ^2 calculation is then

$$\chi^2 = (N_{Rt+1} - E_{Rt+1})^2/(E_{Rt+1})$$
$$+ [(N_{Tt+1} - N_{Rt+1}) - (N_{Tt+1} - E_{Rt+1})]^2/(N_{Tt+1} - E_{Rt+1})$$
$$= (37)^2/(589) + (-37)^2/(1454) = 3.266.$$

Observed numbers of offspring in the two phenotypic classes do not differ significantly from numbers expected on the basis of parental frequencies. Hence allele frequencies in the two generations appear to be approximately equal, and we can use both generations to estimate q.

The frequency of recessives across both generations is

$$N_R = (461 + 626)/(1600 + 2043) = 0.2984,$$

so that

$$\hat{q}' = (N_R)^{0.5} = 0.5462.$$

We can then estimate S_1 as

$$S_1 = \hat{q}'/(1 + \hat{q}') = 0.3533,$$

which we multiply by the total number of children from taster × nontaster marriages, 761, to get

$$E_{S1} = 268.9,$$

the expected number of nontaster children from such marriages. The corresponding observed number was $N_{S1} = 278$, so that

$$\chi_1{}^2 = (9.1)^2/(268.9) + (-9.1)^2/(492.1) = 0.4762.$$

We can most easily estimate S_2 by squaring our estimate of S_1:

$$S_2 = (S_1)^2 = 0.1248.$$

With 1059 total children from taster x taster marriages, we expect 132.2 nontaster children. We observed 130, so that

$$\chi_2{}^2 = (-2.2)^2/(132.1) + (2.2)^2/(926.9) = 0.0419.$$

Neither χ^2 value is greater than the critical value, 3.841. We therefore can conclude, as Snyder did, that the children's genotypes appear in the Hardy-Weinberg ratio.

Nontaster × nontaster marriages contribute nothing to the test for the Hardy-Weinberg ratio. Such marriages will produce only nontaster children, regardless of the parental genotypic ratio. If we write the proportion of nontaster children from such marriages as

$$S_0 = q^0/(1 + q)^0 = 1$$

(any quantity raised to the 0-th power is 1), we see that Snyder's ratios form a progression. With i = 0, 1, or 2, this progression has the general term

$$S_i = q^i/(1 + q)^i, \tag{4.3}$$

where i is the number of dominant parents in the class of matings to which S_i applies.

This raises the speculation that this progression might be extended to higher values of i. Because we only recognize two parents for an individual offspring, the situation to which such an extension might apply is not clear. Doolittle (1968) extended Snyder's ratio calculations to the case of epistasis, which can be regarded as a sort of interlocus dominance. Extension of Eq. (4.3) to $i>2$ did not seem to apply in this situation. It also does not seem to apply to the ratio of Recessive grandchildren from phenotypically classified grandparents, nor to polyploidy.

The natural extension of Snyder's ratios seems to be to a somewhat recondite situation called complementary epistasis (Crow and Kimura 1970). Suppose that an abnormal phenotype is produced by homozygosity for recessive alleles at two loci, say *xx yy*; all other genotypes at the two loci produce normal phenotypes. Then the ratio of abnormal offspring from normal x abnormal matings to those from normal x normal matings turns out to be

$$[q/(1+q)]: [q^2/(1+q)^2],$$

where q is the frequency of *xy* gametes.

The relationship can be extended to more than 2 loci; e.g., when the abnormal phenotype results from the recessive homozygote at three loci, *xx yy zz*. The principle applies even if the loci are linked and in recombination disequilibrium. In essence, Snyder's ratios do not distinguish between inheritance by recessive homozygosity at a single locus and complementary epistasis at two or more loci. This does not, however, answer the question as to the possible meaning of an i value greater than 2.

Snyder's ratios are not the only way to treat the problem of determining whether, in the presence of dominance, a population produces offspring genotypes in the Hardy-Weinberg ratio. Li (1976) reviews a number of alternate approaches to this problem.

Lecture 4 Exercise

A population of mice is segregating for the *B* and *b* alleles at the brown locus in mice. *BB* and *Bb* mice have black pigment in their pelage; *bb*, brown pigment. The following data was obtained in one generation of this population, where it was possible to identify the parents of each mouse:

Parents	Offspring		
	Black	Brown	Total
Black × black	619	45	664
Black × brown	104	50	154
Brown × brown	0	25	25
Totals	723	120	843

Determine whether the offspring display the Hardy-Weinberg ratio.

Part II Constant Allele Frequencies

Lecture 5 Multiple Alleles

The Hardy-Weinberg law states that, under certain assumed conditions, a population carrying alleles A and a at frequencies p and q, respectively, will produce genotypes in the ratio p^2 AA: 2pq Aa: q^2 aa after a single generation of random mating. This is an equilibrium distribution, which will be maintained in each subsequent generation as long as conditions remain the same. The assumptions leading to a Hardy-Weinberg equilibrium are that alleles A and a occur at an autosomal locus in an infinitely large random mating population of a diploid organism, with constant allele frequencies.

We will now begin considering the consequences of making different assumptions about the population. For example, we will shortly consider the consequences of having more than two alleles. As we change assumed conditions, we will in general change only one assumption at a time; e.g., we will consider more than two alleles "at an autosomal locus in an infinitely large random mating population of a diploid organism with constant allele frequencies"; only the number of alleles will be changed. It will later be convenient to consider two or more changes in conditions simultaneously, to observe interactions; we will consider, for example, more than two alleles in a polyploid.

In discussing the effects of a change in conditions, we will first define the new conditions carefully, noting any resulting alteration in the general nature of the population; with multiple alleles, for instance, more than three genotypes can occur. We will then attempt to determine whether an equilibrium is possible under the changed conditions and whether and how the population approaches equilibrium.

We will first consider changes in conditions which do not affect the constancy of allele frequencies, beginning with the introduction of multiple alleles. Green (1966) lists some ten alleles at the agouti locus, and Silvers (1979) a dozen alleles at the pink-eye dilution locus, in mice. These may be extreme cases, but at least 40 of the approximately 300 named loci in the mouse have been shown to carry more than two alleles. Hubby and Lewontin (1966) found multiple alleles at 6 of

Table 5.1. Genotypic distribution for multiple alleles

	A_1 p_{it}	A_2 p_{2t}	$\cdots$	A_n p_{nt}
A_1 p_{1t}	$p_{1t}^2 A_1A_1$	$p_{1t}p_{2t} A_1A_2$	$\cdots$	$p_{1t}p_{nt} A_1A_n$
A_2 p_{2t}	$p_{1t}p_{2t} A_1A_2$	$p_{2t}^2 A_2A_2$	$\cdots$	$p_{2t}p_{nt} A_2A_n$
$\vdots$ $\vdots$	$\vdots$	$\vdots$		$\vdots$
A_n p_{nt}	$p_{1t}p_{nt} A_1A_n$	$p_{2t}p_{nt} A_2A_n$	$\cdots$	$p_{nt}^2 A_nA_n$

21 loci they investigated for electrophoretic variants in Drosophila. Kimura (1983) cites the prevalence of multiple allelism as evidence for his neutral theory of evolution. Clearly, an extension of the Hardy-Weinberg law to the case of multiple alleles is needed. Weinberg recognized this and provided such an extension in his 1909 paper.

Suppose, then, that at some autosomal locus we have, not two alleles, *A* and *a*, but n alleles, A_1, A_2, ..., A_n. In generation t, these alleles occur in the relative frequencies p_{1t}, p_{2t}, ..., p_{nt}, respectively; $\Sigma p_{it}=1$. Because each gamete carries one allele, gamete frequencies are equal to allele frequencies. The Punnett square in Table 5.1 shows the frequencies of offspring genotypes obtained by random union among these gametes.

The cells on the leading diagonal (top left to bottom right) of the table give the frequencies of homozygotes, A_1A_1 through A_nA_n; the frequency of each is the square of the frequency of the allele involved, p_{it}^2. The off-diagonal cells, which give heterozygote frequencies, are symmetrical; the genotype whose frequency appears in row i, column j, is the same as that whose frequency appears in row j, column i. The frequency for each cell is the product of the frequencies of the two alleles in that genotype, $p_{it}p_{jt}$. Because there are two cells with equal frequencies for each heterozygous genotype, the frequency of genotype A_iA_j is $2p_{it}p_{jt}$. In summary, the distribution of genotypes in t+1 is given by the terms of

$$\begin{aligned} &p_{1t}^2+p_{2t}^2+\ldots+2p_{1t}p_{2t}+2p_{1t}p_{3t}+\ldots. \\ &\quad=\Sigma p_{it}^2+2\Sigma p_{it}p_{jt}=(\Sigma p_{it})^2, \end{aligned} \tag{5.1}$$

Σp_{it}^2 representing the frequencies of homozygotes and $2\Sigma p_{it}p_{jt}$ those of heterozygotes.

The frequency of the i-th allele in t+1 will be the frequency of the homozygote A_1A_1, plus half the frequency of any heterozygote in which A_1 occurs. Thus, the frequency of this allele in generation t+1 will be

$$\begin{aligned} p_{1t+1}&=p_{1t}^2+p_{1t}p_{2t}+p_{1t}p_{3t}+\cdots+p_{1t}p_{nt} \\ &=p_{1t}(\Sigma p_{it})=p_{1t}. \end{aligned}$$

Allele frequencies are constant; therefore, gamete frequencies will also be constant, and the genotypic ratio in generation t+2 will be the same as in t+1. Equilibrium exists and is achieved in a single generation of random mating.

From Eq. (5.1), the genotypic distribution at equilibrium can be written as

$$(\Sigma p_i)^2=(p_1+p_2+\ \ldots\ +\ p_n)^2. \tag{5.2}$$

(We can drop the generation index t, because allele frequencies are constant.) We have already seen that, with two alleles, the Hardy-Weinberg ratio is

$$p^2+2pq+q^2=(p+q)^2,$$

and this is $(\Sigma p_i)^2$ for i=1, 2. Two alleles are simply a special case of n alleles, where n=2, A_1 is *A*, A_2 is *a*, and p_1 and p_2 are p and q, respectively.

Lecture 5 Exercises

1. The human ABO blood groups are inherited by three alleles, L^A, L^B, and L^O. Suppose that in a given population the frequencies of these alleles are, respectively, 0.2, 0.1, and 0.7. What will be the equilibrium frequencies of the four blood groups? (Blood group O is $L^O L^O$; group A, $L^A L^A$ or $L^A L^O$; group B, $L^B L^B$ or $L^B L^O$; and group AB, $L^A L^B$.)

2. Three of the alleles at the agouti locus in mice are the agouti (A), black-and-tan (a^t) and nonagouti (a) alleles. What will be the equilibrium frequencies of genotypes and phenotypes in the descendants of a three way cross of stocks (A x B)F x C, if stock A is homozygous AA, stock B homozygous $a^t a^t$, and stock C homozygous aa? (Order of dominance: A, a^t, a.)

Lecture 6 Separate Sexes

In deriving the Hardy-Weinberg law, we assumed that any two gametes could unite to form a new individual. All gametes were drawn from a single pool. The probability that a gamete carried *A* was therefore the same for every gamete; if gametes were drawn at random, this probability was p, the frequency of *A* in the pool.

In higher organisms, however, two types or sexes of gamete are formed, and successful gametic union can only take place between opposite sex gametes. The two sexes of gamete do not come from the same gamete pool, but from separate male and female gamete pools. In most higher plants, however, the same individual produces both male and female gametes. Thus, both male and female gamete pools are necessarily derived from the same array of parent genotypes. Under these circumstances, both gamete pools can be assumed to have the same allele frequencies.

But in some higher plants and in almost all higher animals, the two sexes are lodged in separate individuals. The gametes of the two sexes are produced by different sets of individuals, who may have different genotypic distributions and produce different allelic frequencies in their gametes.

Table 6.1 is a modified version of Table 1.1, which allows for different genotypic distributions in the two sexes in generation t;

D_{mt}: $2H_{mt}$: R_{mt} in males, and

D_{ft}: $2H_{ft}$: R_{ft} in females.

Table 6.1. Derivation of the Bruce distribution

Matings	Frequency	Offspring		
		AA	*Aa*	*aa*
AA × *AA*	$D_{mt}D_{ft}$	$D_{mt}D_{ft}$		
AA × *Aa*	$2D_{mt}H_{ft}$	$D_{mt}H_{ft}$	$D_{mt}H_{ft}$	
AA × *aa*	$D_{mt}R_{ft}$		$D_{mt}R_{ft}$	
Aa × *AA*	$2H_{mt}D_{ft}$	$H_{mt}D_{ft}$	$H_{mt}D_{ft}$	
Aa × *Aa*	$4H_{mt}H_{ft}$	$H_{mt}H_{ft}$	$2H_{mt}H_{ft}$	$H_{mt}H_{ft}$
Aa × *aa*	$2H_{mt}R_{ft}$		$H_{mt}R_{ft}$	$H_{mt}R_{ft}$
aa × *AA*	$R_{mt}D_{ft}$		$R_{mt}D_{ft}$	
aa × *Aa*	$2R_{mt}H_{ft}$		$R_{mt}H_{ft}$	$R_{mt}H_{ft}$
aa × *aa*	$R_{mt}R_{ft}$			$R_{mt}R_{ft}$
	1	$p_{mt}p_{ft}$	$p_{mt}q_{ft}+q_{mt}p_{ft}$	$q_{mt}q_{ft}$

Allelic frequencies may differ correspondingly, with

$$p_{mt} = D_{mt} + H_{mt},\ p_{ft} = D_{ft} + H_{ft},\ \text{etc.}$$

Of course, genotypic and allelic frequencies must each sum to 1 within each sex.

In Table 6.1, the first parent listed for each mating is the male, the second the female. Mating frequencies are simply the products of the appropriate sex-specific genotypic frequencies. The total of the mating frequencies, then, is

$$\begin{aligned} &D_{mt}D_{ft} + 2D_{mt}H_{ft} + D_{mt}R_{ft} + 2H_{mt}D_{ft} + 4H_{mt}H_{ft} + 2H_{mt}R_{ft} \\ &\quad + R_{mt}D_{ft} + 2R_{mt}H_{ft} + R_{mt}R_{ft} \\ &\quad = (D_{mt} + 2H_{mt} + R_{mt})(D_{ft} + 2H_{ft} + R_{ft}) = 1. \end{aligned}$$

The frequencies of *AA*, *Aa*, and *aa* offspring produced are

$$\begin{aligned} D_{t+1} &= D_{mt}(D_{ft} + H_{ft}) + H_{mt}(D_{ft} + H_{ft}) \\ &= (D_{mt} + H_{mt})(D_{ft} + H_{ft}) = p_{mt}p_{ft}; \\ 2H_{t+1} &= p_{mt}q_{ft} + q_{mt}p_{ft}; \\ R_{t+1} &= q_{mt}q_{ft}. \end{aligned} \tag{6.1}$$

This genotypic ratio was first presented by Bruce (1910), and we shall refer to it as the Bruce ratio.

We have assumed that the A locus is autosomal, i.e., independent of the sex determining mechanism. As a result, the probability that a zygote in generation $t+1$ is male will be the same for all zygotes; let this probability be y $(0<y<1)$. Then the frequency of male *AA* zygotes in generation $t+1$ will be yD_{t+1}; of *Aa*, $2yH_{t+1}$. When we take the genotypic ratio, the common factor y will cancel:

$$yD_{t+1}\!: 2yH_{t+1}\!: yR_{t+1} = D_{t+1}\!: 2H_{t+1}\!: R_{t+1}.$$

Similarly, the frequency of *AA* among females in $t+1$ will be $(1\text{-}y)D_{t+1}$, etc., and the common factor 1-y will cancel when we take genotypic ratios. The Bruce ratio applies to both sexes in generation $t+1$, and the relative frequencies of each genotype are the same in both sexes.

But if we have the same genotypic ratio in both sexes in generation $t+1$, Table 6.1 reduces to Table 1.1 for that generation. Allele frequencies must be the same in gametes of both sexes. Thus, in generation $t+2$ the population reaches an equilibrium genotypic ratio, with $D_{t+2} = p_{t+1}^2$. Of course, if genotypic ratios are the same in both sexes of the generation t, equilibrium is achieved in generation $t+1$. Separation of the sexes has no effect on the attainment of equilibrium if both have the same genotypic ratio.

It is the frequency of alleles, rather than that of genotypes, that determines genotypic frequencies in the offspring. Therefore, if $p_{mt} = p_{ft}$, the Bruce ratio reduces to $p_t^2\!: 2p_tq_t\!: q_t^2$ in generation $t+1$, even if the genotypic ratios differ in the two sexes. If, for example,

$$D_{mt} = D_{ft} + x,\ 2H_{mt} = 2H_{ft} - 2x,\ R_{mt} = R_{ft} + x,$$

so that

$$p_{mt} = D_{mt} + H_{mt} = D_{ft} + x + H_{ft} - x = p_{ft},$$

the columns in Table 6.1 will sum to the Hardy-Weinberg ratio; the population will reach equilibrium in generation $t+1$.

If genotypic ratios in the two sexes differ in such a manner that $p_{mt} \neq p_{ft}$, the Bruce distribution does not reduce immediately to the equilibrium distribution. The frequency of *A* in $t+1$ is

$$\begin{aligned} p_{t+1} &= p_{mt}p_{ft} + (p_{mt}q_{ft} + q_{mt}p_{ft})/2 \\ &= (p_{mt}p_{ft} + p_{mt}q_{ft} + p_{mt}p_{ft} + q_{mt}p_{ft})/2 \\ &= (p_{mt} + p_{ft})/2 = \bar{p}_t. \end{aligned} \tag{6.2}$$

Not unreasonably, allele frequencies in generation $t+1$ are the average of those in the two sexes in generation t.

Because allele frequencies are the same in both sexes in generation $t+1$, in $t+2$ the genotypic distribution will be $\bar{p}_t^2 : 2\bar{p}_t\bar{q}_t : \bar{q}_t^2$. Allele frequencies in $t+2$ are again $\bar{p}_t$, $\bar{q}_t$, so that the genotypic distribution in $t+3$ will be the same as in $t+2$, etc. The population has achieved the equilibrium distribution by generation $t+2$.

Equilibrium is not, however, attained in generation $t+1$. The frequency of *AA* in $t+1$, if $p_{mt} \neq p_{ft}$, is

$$p_{mt}p_{ft} \neq \bar{p}_t^2 = [(p_{mt} + p_{ft})/2]^2.$$

One generation is required to render the allele frequencies equal in the two sexes; genotypic equilibrium then follows in the succeeding generation.

Thus, inequality between the sexes is ephemeral, and we are unlikely to find it in any natural population. If, however, we cross two populations, taking males from one and females from the other, the male and female populations may well have different genotypic and allelic frequencies. The Bruce ratio will then represent genotypic frequencies in the F_1 generation; equilibrium will be achieved in F_2.

Indeed, if we cross two populations, X and Y, which have different allelic frequencies ($p_X \neq p_Y$), the genotypes in F_1 will be in the Bruce ratio even if we take individuals of both sexes from either population. Bruce actually developed his ratio in terms of a cross between populations, and did not specify that males were drawn from one, females from the other, population. As long as every mating involves one random mate from each population, the Bruce ratio will apply to the F_1 generation. In F_2, of course, equilibrium will be reached at the ratio $\bar{p}^2 : 2\bar{p}\bar{q} : \bar{q}^2$, $\bar{p} = (p_X + p_Y)/2$.

Suppose, however, that we were to proceed by drawing and pairing mates at random from the two populations, so that any individual from population X would be equally likely to be mated to another from X or to one from Y. Half the matings would then be between one mate from population X and one from population Y. The Bruce ratio would apply to the offspring of these matings. But a fourth of the matings would be between two X individuals, and a fourth between two Y's. These matings would give the Hardy-Weinberg ratio of offspring genotypes from population X or Y, respectively. Then

$$D_{F1} = p_X{}^2/4 + p_X p_Y/2 + p_Y{}^2/4$$
$$= (p_X + p_Y)^2/4 = [(p_X + p_Y)/2]^2 = \bar{p}^2.$$

Equilibrium is reached in a single generation. By randomly choosing mates from both populations, we have, in effect, merged X and Y in generation t into a new population, XY, consisting of equal numbers of individuals from the two original populations. In this new population,

$$p_{XYt} = D_{XYt} + H_{XYt} = (D_{Xt} + D_{Yt})/2 + (H_{Xt} + H_{Yt})/2$$
$$= (D_{Xt} + H_{Xt})/2 + (D_{Yt} + H_{Yt})/2$$
$$= (p_{Xt} + p_{Yt})/2 = \bar{p}.$$

The frequencies of genotypes in XY will not be in the Hardy-Weinberg ratio in generation t:

$$D_{XYt} = (D_{Xt} + D_{Yt})/2 = (p_X{}^2 + p_Y{}^2)/2 \neq [(p_X + p_Y)/2]^2;$$

but equilibrium will be attained in generation t + 1.

Lecture 6 Exercises

1. We cross males from one population with females from a second. The genotypic distributions are 0.4: 0.4: 0.2 among males, 0.2: 0.8: 0 among females. What are the genotypic ratios in the F_1, F_2, and F_3 generations from this cross?

2. Repeat your calculations, assuming now that the male distribution is 0.09: 0.10: 0.81, the female, 0.36: 0.15: 0.49. What are the F_1, F_2, and F_3 genotypic ratios?

3. Suppose that we cross two populations, X and Y, choosing for each mating a male from X and a female from Y, or a male from Y and a female from X. Population X has the same distribution (in both sexes) as the male parents in Exercise 2; population Y, that of the females in that exercise. How does this affect the ratios of genotypes in generations F_1, F_2, and F_3?

4. Now suppose that we choose mates completely at random from the two populations, so that some matings involve two mates from X, some two from Y, and some one mate from each population. What are the genotypic ratios in F_1, F_2, and F_3?

Lecture 7 Sex Linkage

In species where sexes are separate, sex is usually under genetic control. Control of sex may be vested in a single gene locus, or in many; if in many, they may be scattered throughout the genome, or concentrated in specific chromosomes.

In mammals, control is concentrated in a specific chromosome pair. Male and female karyotypes differ. The female has an even number of chromosomes, which pair during meiotic prophase. The two members of each pair are alike in morphology: size, centromere location, banding, etc. The male has a matching or homologous set of pairs, except that he has only a single chromosome corresponding to one of the pairs from the female, usually a medium to large sized metacentric or submetacentric chromosome. At meiosis in the male, this single chromosome pairs with a much smaller chromosome for which the female has no match. The larger male chromosome and its pair of homologues in the female are called the X chromosomes, the smaller male chromosome, the Y. Sex determining genes appear to be concentrated in these chromosomes, so that an XX individual will be a female, an XY a male. Other taxa also show XY sex determination, notably Dipteran (flies) and Coleopteran (beetles) insects. Morgan (1910a) first pointed out the role of this XY mechanism in determining sex in the fruit fly, *Drosophila.*

The two X's of the female pair in meiosis, and all female gametes (ova) carry an X chromosome; the female is said to be homogametic. The male, on the other hand, is heterogametic; half his gametes (sperm) receive an X and half a Y. When an X-bearing ovum unites with a sperm bearing an X, the offspring will be female, XX; but if the sperm carries a Y, the resulting XY zygote will form a male. Thus, the sperm determines the sex of the offspring.

The X chromosome, being larger, carries some gene loci which do not occur on the Y. Such loci are said to be sex-linked. Females carry two copies of the genes at a sex-linked locus, and can be homozygous for any allele, or heterozygous for any pair of alleles, just as for any locus on an autosome (a chromosome other than X or Y). But males carry only one gene at a sex-linked locus. They cannot be either heterozygous or homozygous; the term hemizygous has been coined for them. Because males are, in effect, haploid for sex-linked traits, while females are diploid, the sexes differ in their inheritance and expression of these traits. Morgan (1910b) first described the inheritance of a sex-linked trait (white eye color in *Drosophila*). His student Bridges (1913, 1914) demonstrated that the peculiar inheritance pattern of these traits stemmed from their association with the X chromosome. Jennings (1916) developed the population genetics consequences of sex-linkage.

The demonstration that sex-linked inheritance in *Drosophila* was due to genes carried on the X chromosome was a major factor in Bridges's (1916) declaration that all genes were carried on chromosomes. The reverse system of sex determination appears in birds (Durham and Maryatt 1908) and in Lepidoptera, that is, moths and butterflies (Doncaster and Raynor 1906). Cock birds carry two Z chromosomes, and are homogametic. Hens carry one Z and one W; they are the heterogametic sex. Genes that occur on Z but not on W are sex-linked. The above description of sex-linkage under the XY mechanism of sex determination, and the discussion which follows of the population genetics of sex-linkage, apply equally to the ZW mechanism, but with the roles of the two sexes reversed.

Let *S* and *s* represent a pair of alleles at a sex-linked locus in a diploid organism with separate sexes and XY sex determination. Three genotypes can occur among females, *SS*, *Ss* and *ss*. If the frequencies of these genotypes are D_{ft}, $2H_{ft}$, and R_{ft}, respectively, in generation t, the frequency of *S* alleles among females is

$$p_{ft} = D_{ft} + H_{ft};$$

this will also be the frequency of ova bearing the *S* allele among gametes formed by generation t females.

Males have only one X chromosome, and only two possible genotypes, *S* and *s*. If the frequencies of these genotypes in generation t are designated, respectively, D_{mt} and R_{mt} (there is no H_{mt}), the male frequencies of *S* and *s* in generation t are

$$p_{mt} = D_{mt},\ q_{mt} = R_{mt}.$$

Half the sperm formed by generation t males will carry the Y chromosome and neither *S* nor *s*. But among the half carrying the X chromosome, p_{mt} will be the frequency of sperm carrying *S*, because *S* males will form only *S* sperm, and *s* males, only *s* sperm.

Each female in generation t+1 will receive two X chromosomes, one from her sire and one from her dam. The ratio of genotypes among these females will be the Bruce ratio:

$$SS: Ss: ss = p_{mt}p_{ft}: p_{mt}q_{ft} + q_{mt}p_{ft}: q_{mt}q_{ft}.$$

Among generation t+1 females, then, the frequency of *S* alleles, from Eq. (6.2), will be

$$p_{ft+1} = (p_{mt} + p_{ft})/2. \tag{7.1a}$$

Males in generation t+1, on the other hand, will receive a Y chromosome from their sires. Their X chromosomes will always come from their dams, so that the ratio of male genotypes in generation t+1 will be

$$S: s = p_{ft}: q_{ft};$$
$$p_{mt+1} = p_{ft}. \tag{7.1b}$$

In any generation, female allele frequencies will be the average of male and female frequencies in the previous generation, while male frequencies will be the female frequencies of the previous generation.

If $p_{mt} \neq p_{ft}$, the sex specific allele frequencies change from one generation to the next: $p_{mt+1} = p_{ft} \neq p_{mt}$. Suppose, however, that in some generation e, male and female allele frequencies are the same:

$$p_{me} = p_{fe}. \tag{7.2}$$

From Eq. (7.1):

$$p_{fe+1}=(p_{me}+p_{fe})/2=2p_{fe}/2=p_{fe};$$

$$p_{me+1}=p_{fe}=p_{me}.$$

Male and female allele frequencies are the same in generation e+1 as they were in generation e. Therefore, genotypic frequencies in e+2 must be the same as in e+1. Furthermore, the sex specific allele frequencies will be the same in e+1 as they were in e, and will remain equal to one another; $p_{me+1}=p_{fe+1}$. They will still be equal in generations e+2, e+3, etc. Genotypic frequencies will remain constant from generation e+1 on. The condition expressed in Eq. (7.2), equal allele frequencies in the two sexes, is the necessary and sufficient condition for genotypic equilibrium.

Can a population not in equilibrium in generation t reach equilibrium in a later generation? Let the difference in allele frequencies between the sexes in generation t be

$$d_t=p_{mt}-p_{ft}. \tag{7.3}$$

At equilibrium, of course, $d_e=0$. The difference between male and female allele frequencies in t+1 is

$$\begin{aligned} d_{t+1}&=p_{mt+1}-p_{ft+1}=p_{ft}-(p_{mt}+p_{ft})/2\\ &=(2p_{ft}-p_{mt}-p_{ft})/2=(p_{ft}-p_{mt})/2\\ &=(-1/2)d_t. \end{aligned}$$

In t+1, the difference is half as large as, and of the opposite sign from, the difference in t.

The relationship is recursive; d_{t+2} is to d_{t+1} as d_{t+1} is to d_t, and so on. Therefore,

$$d_{t+2}=(-1/2)d_{t+1}=(1/4)d_t;$$

$$d_{t+3}=(-1/2)d_{t+2}=(-1/8)d_t;$$

$$d_{t+n}=(-1/2)^n d_t. \tag{7.4}$$

Because $(1/2)^n$ goes to 0, d_{tn} goes to 0, as n increases. A population not in equilibrium in generation t will indeed approach equilibrium in later generations.

If the absolute value of the difference between male and female allele frequencies is halved in each successive generation, it will rapidly become very small indeed. In theory, of course, d_{t+n} will approach zero asymptotically as n increases and will never become zero within a finite number of generations. But in any practical sense, d_{t+n} will quickly become so small that it can be ignored. Suppose, for example, that in generation t all males are *S*, all females *ss*. Then $p_{mt}=1$, $p_{ft}=0$, and $d_t=1$, the maximum value it can assume. In successive generations, d will be −0.5, 0.25, −0.125, 0.0625, etc. By generation t+5, the difference will be less than 0.05; by t+7, less than 0.01. Even starting with the maximum difference, d_t becomes negligible within a few generations. The population does not reach equilibrium immediately, but reaches virtual equilibrium very quickly.

Finally, consider the average frequency of *S* genes over both sexes in generation t. Since there are twice as many X chromosomes in females as in males, we must use a weighted average, the frequency in females being given twice the weight of that in males:

$$\bar{p}_t = (p_{mt} + 2p_{ft})/3. \tag{7.5}$$

In generation t+1,

$$\bar{p}_{t+1} = [p_{ft} + 2(p_{mt} + p_{ft})/2]/3$$
$$= (p_{mt} + 2p_{ft})/3 = \bar{p}_t.$$

The weighted average allele frequency over both sexes does not change from one generation to the next. No *S* genes are being created, none are being destroyed; they are merely being shuffled back and forth between males and females. This accounts for the continually reversing sign of d_t.

Because this is true in every generation, it must be true also at equilibrium:

$$\bar{p}_e = (p_{me} + 2p_{fe})/3 = 3p_{me}/3 = p_{me} = p_{fe} = \bar{p}.$$

At equilibrium, the frequency of *S* in either sex is the weighted average frequency, $\bar{p}$.

In the above discussion, we have followed the frequency of the *S* allele, p. Since $p+q=1$ at all times, changes take place in the *s* allele frequency which are equal in amount and opposite in sign to those which take place in p. Only one allele need be followed.

S males carry only one *S* gene, while homozygous *SS* females carry two. Yet, in general, these two genotypes will have the same phenotype; there is seldom any evidence of a dosage effect. Lyon (1961, 1962) hypothesized that only one of the two X chromosomes in any female cell is active. Which chromosome is active is a matter of chance; in *Ss* females, *S* is active in approximately half the cells, *s* in the other half. Usually, the half of the cells in which the *S* is active are sufficient to confer on heterozygotes the dominant phenotype. However, Hulbert and Doolittle (1971), working with a sex linked gene affecting hair growth in mice, found that heterozygous females varied in phenotype. Some showed the dominant phenotype, while others approached the phenotype of the recessive homozygote, with the gamut of intermediate types being present. This was presumably due to variation in the proportion of cells in tissues directly affected by the mutation in which dominant or recessive alleles were active.

Lecture 7 Exercise

The trait rapid-feathering in domestic fowl is controlled by a pair of sex-linked alleles, *K* and *k* (Hutt, 1949). Slow feathering (*K*) is dominant to rapid feathering (*k*). Rapid-feathering Leghorn males are crossed with slow-feathering Rhode Island Red females. What will be the equilibrium ratio of rapid and slow feathering in each sex? Show the expected approach of the population to equilibrium by calculating successive values of $p_f - p_m$ for the first ten generations. (Remember that in fowl, the male is homogametic, the female heterogametic.)

Lecture 8 Other Phenomena Associated with Sex

Partial Sex-Linkage

From voluminous cytogenetic evidence, the pairing of chromosomes at meiotic prophase depends on the pairing of homologous gene loci. Because the X and Y chromosomes do pair at meiosis in males, they can be presumed to carry some gene pairs in common. Sex-linked genes are carried on the X, but not on the Y chromosome; genes that are carried on both the X and the Y chromosomes are said to be "partially sex-linked."

Partially sex-linked loci are diploid in both sexes, as autosomal loci are. But because males pass on their X chromosomes only to their daughters, and their Y's only to their sons, the transmission of traits controlled by partially sex-linked genes will differ from that of autosomally inherited traits (Bennett 1963).

Suppose that we have a pair of alleles, *G* and *g*, at a partially sex-linked locus. A heterozygous *Gg* female will form equal numbers of *G* and *g* ova, so that half her offspring of either sex will receive *G*, and half *g*. But the heterozygous male will pass on one allele to all of his sons, the other to all of his daughters (assuming that there is no crossing-over between the X and Y chromosomes).

In the population genetics of a partially sex-linked trait, then, we must keep track of p_{fX}, the frequency of *G* in females; p_{mX}, that of *G* on the X chromosomes in males; and p_{mY}, that of *G* on Y chromosomes. Frequencies of alleles on Y chromosomes are constant:

$$p_{mYt+1} = p_{mYt} = p_{mY}.$$

As in a sex-linked trait, allele frequencies on the X chromosome vary:

$$p_{fXt+1} = (p_{mXt} + p_{fXt})/2;$$

$$p_{mXt+1} = p_{fXt}.$$

Therefore, as with sex-linkage, if the frequencies of *G* on X chromosomes of the two sexes are equal in some generation e,

$$p_{mXe} = p_{fXe} = \bar{p}_X,$$

frequencies in generation e+1 and all subsequent generations are the same as they were in generation e. If an initial inequality exists between p_{mXt} and p_{fXt}, it tends to dissipate rapidly.

For an autosomal trait, genotypic distributions and allelic frequencies are identical in both sexes at equilibrium. The partially sex-linked trait reaches equilibrium when allele frequencies on X chromosomes are the same in both sexes;

but unless

$$p_{mY} = \bar{p}_X = (p_{mX,t} + 2p_{fX,t})/3,$$

males and females may have different allelic or genotypic frequencies at equilibrium. Among females, the Hardy-Weinberg ratio will prevail, but among males, we will have the Bruce ratio:

$$p_{mY}\bar{p}_X\colon p_{mY}\bar{q}_X + q_{mY}\bar{p}_X\colon q_{mY}\bar{q}_X.$$

Despite the inequality, both $\bar{p}_X$ and p_{mY} are constant from generation to generation, so that the genotypic distributions in the two sexes will also remain constant. (We are still assuming no crossing-over between the X and Y chromosomes.)

Suppose, for example, that we cross males from a population homozygous for *G* with females from a population homozygous for *g*. Then in generation t,

$$p_{fXt} = 0,$$

$$p_{mXt} = p_{mYt} = 1.$$

In t+1,

$$p_{fXt+1} = (p_{mXt} + p_{fXt})/2 = 1/2,$$

$$p_{mXt+1} = p_{fXt} = 0, \text{ and}$$

$$p_{mYt+1} = p_{mYt} = 1.$$

When allele frequencies on the X chromosome reach equilibrium,

$$p_{fXe} = p_{mXe} = \bar{p}_X = 1/3,$$

but p_{mYe} remains constant at 1. If *G* is dominant to *g*, all males will be of the dominant phenotype, but four females in nine will be recessive. The population will be in equilibrium, if there is no crossing over between X and Y chromosomes, despite the inequality between the sexes.

If crossing over can occur between the X and Y chromosomes, of course, Y chromosomes carrying the *g* allele will be formed. Then p_{mYt} will not be constant, but will change slowly, the rate of change being determined by the rate of crossing over between X and Y. Eventually,

$$p_{fXe} = p_{mXe} = p_{mYe}.$$

In the case of the cross between two homozygous populations described above, this frequency will be 1/2. The Hardy-Weinberg genotypic ratio will then apply across both sexes.

At one time, a number of human traits were ascribed to partial sex-linkage (Lenz 1961). A more recent medical genetics text (Thompson and Thompson 1980) does not even mention the term. Apparent sex inequalities in these traits have either been shown to be spurious, or have been explained on the basis of other modes of inheritance. The bobbed gene in *Drosophila*, however, does appear to be a true example of this type of inheritance (Stern 1926, 1927).

Holandry

Partially sex-linked genes are carried by both the X and the Y chromosomes. The X carries sex-linked genes, which are not carried on the Y. It is logically symmetrical to suppose that the Y chromosome carries genes not carried by X. The term "holandric" is used to refer to such genes. (In ZW sex determination, the corresponding W-only genes are called "hologynic".)

Figure 8.1 provides a diagrammatic representation of the X and Y chromosomes. This shows the locations of sex-linked, holandric, and partially sex-linked genes relative to homologous and nonhomologous segments of the chromosomes.

A holandrically inherited abnormality will show certain transmission characteristics. It will occur, first of all, only in males, never in females. All male descendants in a male line from an afflicted individual will also be affected, but a female descendant will neither show the abnormality, nor pass it on to her descendants; even her sons will not have the abnormality unless they inherit it from their sire(s). Every male has only one gene for this trait, and dominance cannot be defined.

The population genetics of holandric inheritance is very simple. The frequency of the abnormal gene is easily estimated from the number of afflicted males divided by the total number of males in the population. Since the gene occurs only on the Y chromosome, each male is counted only once, and females are not counted at all. In the absence of mutation or selection, the frequencies of holandric genes are constant.

At one time (Winchester 1951), a number of human traits were reported to show holandric inheritance. Stern (1957) examined the pedigrees for 17 such traits, and concluded that there was no compelling evidence for Y-linkage in any

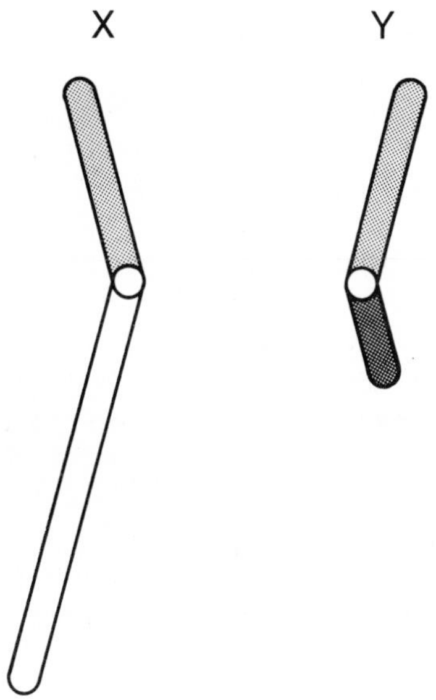

Fig. 8.1. The X and Y chromosomes; homologies and differences. The *unshaded* long arm of the X chromosome carries the sex-linked genes, not homologous to Y. The *heavily shaded* short arm of Y is not homologous to X; genes in this arm are holandric. Genes in the *lightly shaded* upper arms of both chromosomes are the homologous partially sex linked genes

of them. Dronamraju (1965), however, has argued fairly convincingly that *hypertrichosis pinnae auris* (hairy ear lobes) in Hindu populations is inherited holandrically. Shortly after Stern's rejection of Y-linkage, however, holandric inheritance was clearly demonstrated for one very important trait: maleness.

Bridges (1916) had studied nondisjunction of sex chromosomes in fruit flies and other species, and had concluded that the Y chromosome was inert in sex determination. In fruit flies, the XO sex chromosome constitution gave rise to males, the XXY to females, albeit in both cases abnormal and sexually immature individuals. Sex was determined by the balance between the number of X chromosomes and the number of sets of autosomes.

Since mammals, man among them, shared the XY mechanism of sex determination with *Diptera*, it was assumed that the mammalian Y chromosome, like that of fruit flies, was inert with regard to sex determination. But in 1959, Ford et al. applied new methods of cytogenetic examination to Turner's syndrome, and Jacobs and Strong to Klinefelter's syndrome.

Turner (1938) had described a syndrome of infantilism and immature development of sex organs and secondary sexual characteristics in human females. When Ford and his colleagues examined the karyotypes of patients with Turner's syndrome, using new techniques that made possible the more accurate determination of chromosome numbers and characteristics, they found that these individuals had only one X chromosome, and no Y. Klinefelter et al. (1942) had described a syndrome of feminization and immature sexual development in human males; Jacobs and Strong showed that this syndrome was associated with the XXY sex chromosome karyotype.

In contrast to fruit flies, humans lacking the Y chromosome were females. Those with a Y were males, even if they carried the normal female complement of X chromosomes as well. These results were soon confirmed in other mammals. It therefore appears that in mammals, the Y chromosome carries a gene or genes for maleness, not paired with X chromosome genes. Maleness in mammals is a holandric trait.

Sex-Limited and Sex-Influenced Traits

Sex-limited traits can only be expressed in one sex. An example of such a trait is hen-feathering in domestic fowl (Hutt, 1949). Males of most breeds grow long, narrow, sharp-pointed feathers on the neck, saddle and back, called cock feathers. Females grow a shorter, broader, round-tipped hen feather in these areas. Both males and females of the Sebright Bantam breed, however, grow only hen feathers. The Sebright Bantam is homozygous for a recessive gene, *hf*, which apparently prevents the formation of cock feathers in *hfhf* males.

Breeds which grow cock feathers in the males are homozygous for the dominant allele, *Hf*. The henfeathering locus is autosomal, and both males and females are homozygous *HfHf*, but cock-feathering cannot be expressed in the female. It is a sex-limited trait, inhibited by the presence of female ovarian hormone.

Similarly, Dawson and Riddle (1975) reported an egg abnormality in *Tribolium* flour beetles, inherited by an autosomal recessive gene. Females homozygous for this gene produce eggs of abnormal phenotype; because males do not produce eggs, they cannot express the gene even if homozygous for it.

The sex-limited inheritance of these traits thus resembles that of other traits of considerably greater importance to animal breeders, which can be described as sex limited; e.g., egg production in fowl and milk production in dairy cattle. These are, of course, quantitative traits, inherited by the action and interaction of a large number of gene loci, probably mostly autosomal. Nonetheless, by their nature they can be expressed only in one sex. A dairy bull, no matter how many genes for high milk production he may carry, will never produce a drop of milk. This presents some difficulty for the dairy breeder; he can only judge the bull's genotype by the production records of his female ancestors, collaterals, and descendants.

A final category of traits associated with sex is that referred to as sex-influenced. An abnormality showing sex-influenced inheritance shows differences in penetrance between males and females, probably because of differences in sex hormones between the two. In extreme cases, the allele which is dominant in one sex may be recessive in the other. Pattern baldness in humans is often quoted as an example of a sex-influenced trait. Heterozygous males will usually express the trait; heterozygous females will not (Thompson and Thompson 1980). In fact, because the gene is not often expressed even in homozygous females unless there is a hormonal imbalance, geneticists now tend to regard this trait as sex-limited.

Both sex-influenced and sex-limited traits are inherited by autosomal genes. Their inheritance, as far as genotypes are concerned, is normal for autosomal genes; only the expression of the genotypes differs between the sexes. The population genetics of these traits is therefore normal for an autosomal trait.

Lecture 8 Exercise

We have a human pedigree of an unusual enzyme abnormality. The proband of this pedigree was a male (I-1) lacking the enzyme atobase. Married to a normal woman, whose blood relatives were all normal, I-1 produced a son, II-1, who also

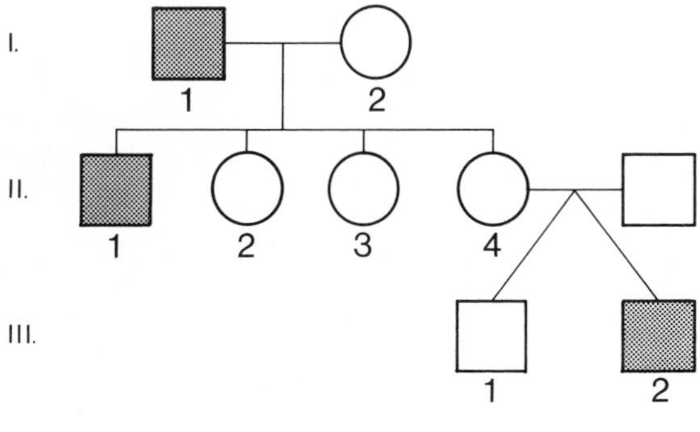

Fig. 8.2. The atobase human pedigree. *Shaded* individuals lack the enzyme atobase

lacked the enzyme, and three daughters, II-2, −3 and −4, all of normal phenotype. II-4, however, produced one son, III-1, who was normal, and a second, III-2, who lacked the enzyme. The husband of II-4 was normal, as were the husbands and all offspring of II-2 and II-3. Is this abnormality most likely to be sex-linked, holandric or sex-limited in inheritance? Explain the rationale behind your answer.

Lecture 9 Linkage

A single autosomal locus in a Hardy-Weinberg population reaches equilibrium in one generation, at genotypic frequencies that are simple functions of the frequencies of the alleles. Suppose, however, that we are interested in genotypes at two different loci at the same time. We may want to predict the frequencies of combinations of traits controlled by two loci; or two loci may both act upon the same trait, the phenotype of an individual being determined by the interaction between genotypes at the two loci. As we shall see later, quantitative traits, of great importance to plant and animal breeders, are inherited by the combined effects of a number of loci.

When we consider two or more loci at the same time, we must consider whether or not the loci are linked. Bateson and Punnett (1906) first described the phenomenon of linkage, noting that in some instances, traits paired in the parents of a cross tended to remain together in the offspring. They referred to the situation when the two dominant traits tended to segregate together into one offspring, and the two recessive traits into another, as "linkage in coupling." When the adherent pairs were dominant-recessive and recessive-dominant, they called it "linkage in repulsion." Unaware of the association of genes with chromosomes, they invented an elaborate mechanism of post-meiotic cell divisions to account for the phenomenon.

Morgan et al. (1915) and Bridges (1916) proposed a linear arrangement of genes along the chromosomes which clarified the cause of linkage (See Fig. 9.1). The integrity of the chromosome tends to cause genes carried on the same chromosome to segregate together. Crossing over, the exchange of segments between sister chromatids permits the recombination of linked genes. The frequency of crossovers that recombine a pair of linked loci depends on the linear distance between the genes along the chromosome. Loci that are farther apart recombine more frequently. Loci on different chromosomes, of course, segregate independently.

In considering the population genetics of a pair of loci, then, we must consider the effect of linkage; Jennings (1917) and Robbins (1918) were the first to assess these effects. It might appear that linkage is a complication in developing the population genetics of two loci considered simultaneously, and that we would be better off to begin with a consideration of unlinked loci, segregating independently. In fact, it turns out to be easier to develop our argument in terms of linked loci, later treating independent segregation as a special case of linkage. We shall therefore assume first that we are dealing with a pair of linked loci.

Consider, then, a random mating population of diploid organisms, in which alleles *A* and *a* are segregating at one autosomal locus, and *B* and *b* at a second. Let allelic frequencies in generation t be p_t for *A*, q_t for *a*, r_t for *B*, and s_t for *b*;

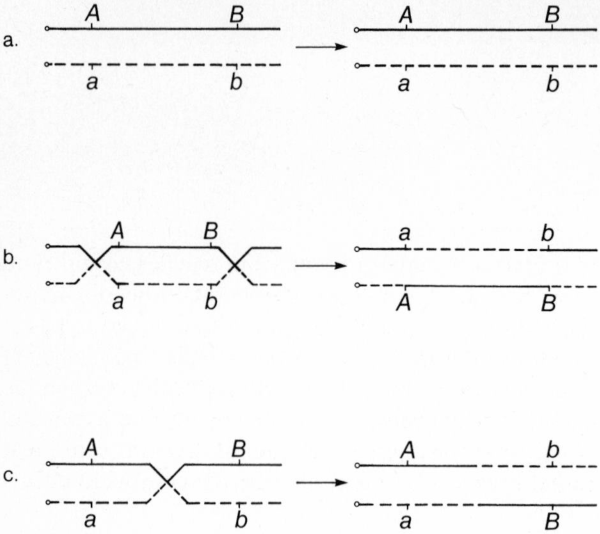

Fig. 9.1. Linkage of the A and B loci. (a)A coupling double heterozygote. The loci are represented as being on a pair of telocentric autosomes. (b)Crossing over outside the chromosome segment between the two loci does not affect recombination; (c)crossing over within that segment does

$p_t + q_t = r_t + s_t = 1$, of course. Let us further suppose that the two loci occur on the same chromosome, and recombine by crossing over with a frequency c per generation.

Any gamete formed in this population must carry one allele from each of the two loci. Therefore, the population will form four types of gamete, *AB*, *Ab*, *aB*, and *ab*. Let g_{ijt} designate the frequency in generation t of gametes carrying the i-th allele at the A locus and the j-th allele at the B locus. If we let $i=1$, for *A*, 2 for *a*, and $j=1$ for *B*, 2 for *b*, the frequency of *AB* gametes in generation t will be g_{11t}; of *Ab* gametes, g_{12t}; of *aB*, g_{21t}; and of *ab*, g_{22t}. The sum of the g_{ijt} is

$$g_{ijt} = g_{11t} + g_{12t} + g_{21t} + g_{22t} = 1.$$

For a single locus, gamete and allele frequencies were the same; this is not true in the case of two loci. Allele and gamete frequencies are, however, related. Gametes carrying *A* must have a total frequency of p_t, etc., so that

$$p_t = g_{11t} + g_{12t};$$

$$q_t = g_{21t} + g_{22t};$$

$$r_t = g_{11t} + g_{21t};$$

$$s_t = g_{12t} + g_{22t}. \qquad (9.1)$$

The genotypic frequencies which result when these gametes are paired at random are shown in Table 9.1. The student can confirm these results, using a 4×4 Punnett square, if he so desires. Two interesting facts can be observed from Table 9.1.

Table 9.1. Genotypic frequencies in generation t+1

	AA	*Aa*	*aa*	Total
BB	g_{11t}^2	$2g_{11t}g_{21t}$	g_{21t}^2	$(g_{11t}+g_{21t})^2$
Bb	$2g_{11t}g_{12t}$	$2g_{11t}g_{22t}+2g_{12t}g_{21t}$	$2g_{21t}g_{22t}$	$2(g_{11t}+g_{21t})(g_{12t}+g_{22t})$
bb	d_{12t}^2	$2g_{12t}g_{22t}$	r_{22t}^2	$(g_{12t}+g_{22t})^2$
Total	$(g_{11t}+g_{12t})^2$	$2(g_{11t}+g_{12t})(g_{21t}+g_{22t})$	$(g_{21t}+g_{22t})^2$	1

First, examine the marginal totals. The total of the second column, for example, is

$$
\begin{aligned}
&2g_{11t}g_{21t}+2g_{11t}g_{22t}+2g_{12t}g_{21t}+2g_{12t}g_{22t}\\
&\quad=2[g_{11t}(g_{21t}+g_{22t})+g_{12t}(g_{21t}+g_{22t})]\\
&\quad=2(g_{11t}+g_{12t})(g_{21t}+g_{22t})=2p_tq_t.
\end{aligned}
$$

The total of this column represents the total frequency of *Aa* offspring, summed across B locus genotypes, and this total, from Eq. (9.1), is $2p_tq_t$. The total of the first column is the frequency of *AA* and this total is $p_t{}^2$; the total of the third column, the frequency of *aa*, is $q_t{}^2$. The A locus genotypes are in Hardy-Weinberg equilibrium after a single generation. Similarly, the row totals from top to bottom are the frequencies of *BB*, *Bb*, and *bb*, and these are $r_t{}^2$, $2r_ts_t$, and $s_t{}^2$, respectively. The B locus is also in Hardy-Weinberg equilibrium after one generation. Each locus, considered by itself, reaches Hardy-Weinberg equilibrium in a single generation, as we would indeed expect that it should.

The second point of interest in Table 9.1 is that there are two terms in the expression for the frequency of *AaBb* double heterozygotes (row 2, column 2). Two types of gametic union can give rise to this genotype, *AB* with *ab*, and *Ab* with *aB*. The frequency of unions of *AB* with *ab* is $2g_{11t}g_{22t}$; that of *Ab* with *aB*, $2g_{12t}g_{21t}$. The two terms in the *AaBb* frequency in Table 9.1 thus record the frequencies of double heterozygotes arising from the two types of gametic union. Borrowing from the language of Bateson and Punnett (1906), we will refer to these as coupling and repulsion double heterozygotes, respectively. Similarly, we will call *AB* and *ab* coupling gametes, and *Ab* and *aB* repulsion gametes.

We must distinguish between the two types of double heterozygote because they produce gametes in different frequencies. Instead of the usual "locus A-locus B genotype" format (*AaBb*), we will use a "gamete/gamete" notation, writing the two gametes that unite to form a given genotype with a slash (/) between them. Thus, the coupling double heterozygote will be written as *AB/ab*, and the repulsion double heterozygote as *Ab/aB*. We will use a similar notation for the other genotypes, e.g., *AABB* as *AB/AB* and *AaBB* as *AB/aB*, even though each such genotype can only result from a single type of gametic union.

This notation makes it easy to determine the gametic production of each genotype. If recombination does not occur, a given genotype will produce the same types of gametes that went into its formation. In recombination, the A locus genes exchange B locus partners. Thus, the nonrecombinant gametes formed by geno-

Table 9.2. Gametes produced by genotypes in generation t+1

Genotype	Frequency	Gametes			
		AB	*Ab*	*aB*	*ab*
AB/AB	g_{11t}^2	1			
AB/Ab	$2g_{11t}g_{12t}$	1/2	1/2		
AB/aB	$2g_{11t}g_{21t}$	1/2		1/2	
AB/ab	$2g_{11t}2_{22t}$	(1−c)/2	c/2	c/2	(1−c)/2
Ab/Ab	g_{12t}^2		1		
Ab/aB	$2g_{12t}g_{21t}$	c/2	(1−c)/2	(1−c)/2	c/2
Ab/ab	$2g_{12t}g_{22t}$		1/2		1/2
aB/aB	g_{21t}^2			1	
aB/ab	$2g_{21t}g_{22t}$			1/2	1/2
ab/ab	g_{22t}^2				1

type *AB/ab* are *AB* and *ab*; the recombinant gametes, *Ab* and *aB*. If recombination occurs with frequency c, the *AB/ab* genotype will produce *AB*, *Ab*, *aB* and *ab* gametes in the proportions (1−c)/2: c/2: c/2: (1−c)/2. The *Ab/aB* genotype, on the other hand, will produce the same array of gametes in the proportions c/2: (1−c)/2: (1−c)/2: c/2.

Recombination only affects the proportions of gametes from double heterozygotes. Double homozygotes form the same type of gametes, whether recombination takes place or not; e.g., *AB/AB* forms only *AB* gametes in either case. Likewise, single homozygotes form the same gamete types regardless of recombination; *AB/Ab* forms equal numbers of *AB* and *Ab* gametes in recombinant and in nonrecombinant gametes. Gamete production in generation t+1 is summarized in Table 9.2.

In Table 9.2, the sum of the entries in the *AB* column, each multiplied by the corresponding genotypic frequency, will give the frequency of *AB* gametes formed in generation t+1. That sum is

$$\begin{aligned} g_{11t+1} &= g_{11t}^{\;2} + g_{11t}g_{12t} + g_{11t}g_{21t} + (1\text{-c})g_{11t}g_{22t} + cg_{12t}g_{21t} \\ &= g_{11t}(g_{11t} + g_{12t} + g_{21t} + g_{22t}) - c(g_{11t}g_{22t} - g_{12t}g_{21t}) \\ &= g_{11t} - c(g_{11t}g_{22t} - g_{12t}g_{21t}). \end{aligned}$$

The frequency of gamete *Ab* is

$$\begin{aligned} g_{12t+1} &= g_{11t}g_{12t} + cg_{11t}g_{22t} + g_{12t}^{\;2} + (1\text{-c})g_{12t}g_{21t} + g_{12t}g_{22t} \\ &= g_{12t}(g_{11t} + g_{12t} + g_{21t} + g_{22t}) + c(g_{11t}g_{22t} - g_{12t}g_{21t}) \\ &= g_{12t} + c(g_{11t}g_{22t} - g_{12t}g_{21t}). \end{aligned}$$

Frequencies of *aB* and *ab* are, respectively,

$$g_{21t+1} = g_{21t} + c(g_{11t}g_{22t} - g_{12t}g_{21t}), \text{ and}$$

$$g_{22t+1} = g_{22t} - c(g_{11t}g_{22t} - g_{12t}g_{21t}).$$

Thus, in generation t+1, the frequency of a given gamete is the frequency of the same gamete in generation t, plus or minus the recombination frequency multiplied by the difference between the product of coupling and the product of repulsion gamete frequencies in generation t.

This product difference, $g_{11t}g_{22t} - g_{12t}g_{21}$, is a common factor in all of the generation t+1 gamete frequencies, and we can let

$$d_t = g_{11t}g_{22t} - g_{12t}g_{21t}. \quad (9.2)$$

Then the t+1 gamete frequencies are

$$g_{11t+1} = g_{11t} - cd_t;$$

$$g_{12t+1} = g_{12t} + cd_t;$$

$$g_{21t+1} = g_{21t} + cd_t;$$

$$g_{22t+1} = g_{22t} - cd_t. \quad (9.3)$$

Thus, cd_t measures the change in gamete frequencies between generations t and t+1. This quantity is added to the frequencies of repulsion gametes, subtracted from those of coupling gametes. c is never negative; d_t can be either positive or negative. Equations (9.3) thus do not imply that coupling gamete frequencies always decrease, repulsion gamete frequencies always increase; if d_t is negative, the opposite changes occur. The pluses and minuses merely show that, whenever coupling gametes increase in frequency, repulsion gametes must decrease, and vice versa.

We have noted above that gamete frequencies are not, in the case of two loci, equal to allele frequencies. Gamete frequencies change, but allele frequencies do not. The relationships between the frequencies of alleles and gametes expressed in Eqs. (9.1) represent a simple counting of alleles in the various gametes. Hence, they hold true in every generation. For example,

$$p_{t+1} = g_{11t+1} + g_{12t+1}.$$

But then

$$p_{t+1} = g_{11t} - cd_t + g_{12t} + cd_t = p_t.$$

The change in the frequency of the coupling gamete included in the *A* allele frequency cancels with the change in the frequency of the repulsion gamete, so that the allele frequency remains the same. *A* (or *a*) alleles are neither created nor destroyed; they are merely transferred from a gamete bearing *B* to one bearing *b*, or vice versa. B locus alleles are similarly transferred between *A* and *a* gametes.

The amount of change in gamete frequencies depends on the recombination rate c, as well as upon d_t. If the A and B loci are not linked, but occur on different chromosomes, we expect the alleles at these two loci to segregate independently. In the coupling double heterozygote, *AB*/*ab*, half the A locus genes are *A*, half *a*. Therefore, half of the gametes produced from this genotype will carry *A*, half *a*. Similarly, half the gametes will carry *B*, half *b*. If the two loci segregate independently, half the gametes carrying *A* will also carry *B*; the frequency of *AB* gametes will be (1/2)(1/2) = 1/4. By similar reasoning, the frequencies of *Ab*, *aB*, and *ab* gametes will each also be 1/4. And the repulsion double heterozygote, *Ab/aB*, will also produce 1/4 of each of the four types of gametes. But *AB/ab* produces the four gametes in frequencies (1-c)/2, c/2, c/2, and (1-c)/2, respectively, from Table 9.2. Therefore, if the two loci segregate independently,

$$(1-c)/2 = c/2 = 1/4,$$

which can only be true if c = 1/2. Independent assortment of genes at the two loci is equivalent to a recombination rate of 1/2.

Loci which are close together on the same chromosome recombine with some frequency less than 1/2. The alleles do not segregate independently into the gametes. If *A* and *B* are on the same chromosome, they will tend to segregate together, and will separate only when recombination occurs due to crossing over between them.

As we consider pairs of loci that are farther and farther apart on the same chromosome, the recombination frequency, c, increases. If the two loci are very far apart, crossing over between them may occur so frequently that they recombine as often as not; c = 1/2. Loci so far apart on the same chromosome thus segregate independently, as if they were on different chromosomes.

The maximum crossover frequency is 1/2, because, as Janssens (1909) speculated and Creighton and McClintock (1931) demonstrated, a single crossover event involves only one of the two daughter chromatids of each of a pair of chromosomes. Operationally, we can regard the maximum value of c in terms of mechanisms. If loci segregate independently, c = 1/2. We know of a mechanism, linkage, which can reduce the frequency of recombination below 1/2. We do not know of any mechanism that will increase this frequency above 1/2.

The above discussion has been in terms of the recombination frequency, c. Although the mechanism of recombination between linked loci is crossing over, c is not necessarily the frequency of crossover events. For example, if two crossovers occur between a pair of loci, one negates the effects of the other, and no recombination occurs. c is 0, even though there have been two crossover events. A fuller discussion of the relationship between the recombination rate and the rate of crossing over is given by Crow and Kimura (1970).

Lecture 9 Exercises

1. The inbred mouse strain BALB/c is homozygous *AAcc* at the agouti and albino loci, respectively. Inbred strain C57BL/6 is homozygous *aaCC*. The agouti and albino loci are in different linkage groups. What are the expected proportions of gametes produced by the F_1 of a cross between these two strains? What are the expected genotypic proportions in F_2?

2. The gamete array in a given population in generation t is

$$g_{11t}=0.13,\ g_{12t}=0.17,\ g_{21t}=0.17,\ g_{22t}=0.53.$$

Calculate the gamete arrays for the next three generations if $c=1/2$, and if $c=1/4$.

Lecture 10 Recombination Equilibrium

We have been considering an infinitely large random mating population carrying two gene loci on an autosome, with recombination occurring between the two loci with frequency c. Gametes produced by this population will each carry one allele from each of the two loci. Under these circumstances, as we have seen [Eqs. (9.3)], cd_t represents the change in gamete frequencies from generation t to generation t+1. Therefore, if in some generation e, $cd_e = 0$, there will be no change in gamete frequencies between generation e and e+1:

$$g_{11e+1} = g_{11e} - cd_e = g_{11e}, \text{ etc.}$$

Then the genotypic distribution in generation e+2 will be the same as in e+1; equilibrium will have been attained. Thus, $cd_e = 0$ is the condition required for recombination equilibrium.

If two loci lie very close together on the same chromosome, recombination between them may occur very rarely; c is 0, or nearly 0. In such a case, we are likely to perceive the two as a single locus, and to regard the different gametic combinations as alleles (psuedoallelism). The recombination rate being of the same order of magnitude as a mutation rate, the rare recombination events that do occur are interpreted as mutations. Fisher suggested (Race, 1944) that the inheritance of human Rh blood types could best be explained as the result of a complex of three very closely linked loci, each with two alleles controlling the presence or absence of one of three specific antigens. Likewise, Wallace (1965) suggested two very closely linked loci associated with the two functions (agouti banding and light belly) controlled by the agouti "locus" in mice.

Aside from this rather uncommon situation, c will always be positive, $0 < c \leqq 1/2$. Then $cd_e = 0$ only if $d_e = 0$. The population will approach equilibrium only if d_{t+n} goes to zero as n increases. From the definition of d_t [(Eq. (9.2)],

$$\begin{aligned} d_{t+1} &= (g_{11t} - cd_t)(g_{22t} - cd_t) - (g_{12t} + cd_t)(g_{21t} + cd_t) \\ &= g_{11t}g_{22t} - g_{12t}g_{21t} - g_{11t}cd_t - g_{22t}cd_t - g_{12t}cd_t - g_{21t}cd_t + (cd_t)^2 - (cd_t)^2 \\ &= d_t - cd_t(g_{11t} + g_{22t} + g_{12t} + g_{21t}) \\ &= d_t - cd_t = (1\text{-}c)d_t. \end{aligned}$$

And because this relationship is recursive,

$$d_{t+2} = (1\text{-}c)d_{t+1} = (1\text{-}c)^2 d_t;$$

$$d_{t+n} = (1\text{-}c)^n d_t. \qquad (10.1)$$

Because c is a fraction between 0 and 1/2, $(1-c)$ is a fraction between 1/2 and 1. Therefore, from Eq. (10.1), d_{t+n} approaches 0 asymptotically as n becomes large, because the fraction $(1-c)^n$ decreases. The population will indeed approach equilibrium.

The recombination equilibrium state has a number of interesting facets, stemming from a basic relationship among gamete frequencies in a population which has reached equilibrium. Because

$$d_e = g_{11e}g_{22e} - g_{12e}g_{21e} = 0,$$

it follows that

$$g_{11e}g_{22e} = g_{12e}g_{21e}. \qquad (10.2)$$

At equilibrium, the product of the frequencies of the coupling gametes is equal to the product of the frequencies of the repulsion gametes.

Because $\Sigma g_{ije} = 1$, we can multiply g_{11e}, the equilibrium frequency of *AB* gametes, by this sum without changing its value:

$$\begin{aligned} g_{11e} &= g_{11e}(g_{11e} + g_{12e} + g_{21e} + g_{22e}) \\ &= g_{11e}^{\;2} + g_{11e}g_{12e} + g_{11e}g_{21e} + g_{11e}g_{22e}. \end{aligned}$$

But from Eq. (10.2), we can substitute $g_{12e}g_{21e}$ for $g_{11e}g_{22e}$:

$$\begin{aligned} g_{11e} &= g_{11e}^{\;2} + g_{11e}g_{12e} + g_{11e}g_{21e} + g_{12e}g_{21e} \\ &= g_{11e}(g_{11e} + g_{12e}) + g_{21e}(g_{11e} + g_{12e}) \\ &= (g_{11e} + g_{12e})(g_{11e} + g_{21e}). \end{aligned}$$

Therefore, from Eqs. (9.1),

$$g_{11e} = pr. \qquad (10.3a)$$

That is to say, the frequency of *AB* gametes at equilibrium is the product of the frequencies of the *A* and *B* alleles. The other equilibrium gamete frequencies can be expressed in a similar manner:

$$g_{12e} = ps;$$

$$g_{21e} = qr;$$

$$g_{22e} = qs. \qquad (10.3b)$$

The frequency of each type of gamete at equilibrium is the product of the frequencies of the alleles in that type of gamete.

From Table 9.1, the frequency of *AB*/*AB* double homozygotes at equilibrium is

$$g_{11e}^{\;2} = p^2r^2,$$

that is, the product of the frequencies of *AA* (p^2) and *BB* (q^2). Similarly, the equilibrium frequency of *AB*/*aB* is

$$2g_{11e}g_{21e} = 2pqr^2.$$

Table 10.1. Genotypic frequencies at equilibrium

	AA	*Aa*	*aa*	Total
BB	p^2r^2	$2pqr^2$	q^2r^2	r^2
Bb	$2p^2rs$	$4pqrs$	$2q^2rs$	$2rs$
bb	p^2s^2	$2pqs^2$	q^2s^2	s^2
Total	p^2	$2pq$	q^2	1

Genotypic frequencies at equilibrium are summarized in Table 10.1. The frequency of a two-locus genotype at equilibrium is the product of the equilibrium frequencies of the genotypes at the two loci.

We can define the difference between the state of the population in any generation t and its equilibrium state as the disequilibrium in generation t. In an actual population, we can observe the frequencies of A locus genotypes, summed across genotypes at the B locus, and of B locus genotypes, summed across the A locus, in generation t. Even if there is dominance at both loci, we can observe the frequencies of *aa* and *bb* genotypes. These frequencies estimate the equilibrium frequencies of the two genotypes, because each locus singly is in equilibrium after one generation of random mating. Thus their product estimates the equilibrium frequency of *ab*/*ab* genotypes, and the square root of their product estimates $g_{22e}=qs$, the equilibrium frequency of *ab* gametes. We can also observe the frequency of double recessive homozygotes in generation t; the square root of this frequency is an estimate of the frequency of *ab* gametes in generation t, g_{22t}. The difference between the generation t and the equilibrium frequencies of *ab* gametes (or of any other gametes for which we can obtain estimates) would seem an appropriate measure of disequilibrium in generation t:

$$\hat{g}_{22t}-\hat{g}_{22e}=\hat{g}_{22t}-\hat{q}\hat{s}.$$

But this difference estimates d_t, because

$$\begin{aligned} g_{22t}-qs &= g_{22t}-(g_{21t}+g_{22t})(g_{12t}+g_{22t}) \\ &= g_{22t}-g_{12t}g_{21t}-g_{21t}g_{22t}-g_{12t}g_{22t}-g_{22t}^{\;2} \\ &= g_{22t}(1-g_{21t}-g_{12t}-g_{22t})-g_{12t}g_{21t} \\ &= g_{11t}g_{22t}-g_{12t}g_{21t}=d_t. \end{aligned} \tag{10.4}$$

Thus, d_t, which can be estimated by the difference between a gamete's frequency in generation t and its expected frequency at equilibrium, measures disequilibrium in generation t. This seems reasonable in view of the fact that d goes to 0 at equilibrium.

If we divide both sides of Eq. (10.2) by $g_{21e}g_{22e}$,

$$(g_{11e}g_{22e})/(g_{21e}g_{22e})=(g_{12e}g_{21e})/(g_{21e}g_{22e});$$

$$g_{11e}/g_{21e}=g_{12e}/g_{22e}. \tag{10.5a}$$

But g_{11e} is the equilibrium frequency of the *AB* gamete, g_{21e} of the *aB* gamete, and g_{11e}/g_{21e} is the ratio of the frequencies of gametes bearing *A* and those bearing *a* among gametes that carry *B*. Equation (10.5a) therefore says that the ratio of *A* to *a* among gametes that carry *B* is the same as the ratio of *A* to *a* among those that carry *b*, at equilibrium.

If we divide both sides of Eq. (10.2) by $g_{12e}g_{22e}$, we obtain

$$g_{11e}/g_{12e} = g_{21e}/g_{22e}; \tag{10.5b}$$

the ratio of *B* to *b* is the same in *A* as in *a* gametes. Equations (10.5) again imply the independent assortment of the two loci at equilibrium.

An interesting consequence of this independence is that we cannot detect linkage between the A and B loci at equilibrium from any association between traits controlled by the two loci in the population as a whole. Whether or not linked, the two loci will come to the same equilibrium point, and at equilibrium, will be independently distributed. There will be no association between the two traits at equilibrium due to linkage in data from the whole population (although, of course, linkage will be detectable in the correlated assortment of traits from specific families).

It should be noted that the disequilibrium remaining in generation $t+1$ is $(1-c)$ times the disequilibrium in generation t. For nonlinked loci, $1-c=1/2$, so that half of the disequilibrium in a given generation will remain in the next. The disequilibrium will disappear rapidly if the loci are not linked, but they will not reach equilibrium immediately. For this reason, we refer to "recombination disequilibrium", rather than "linkage disequilibrium"; disequilibrium can persist, for a few generations at least, even between loci that are not linked.

Consider two populations, X and Y, in which the A and B loci are segregating; the recombination rate between the two loci is c in each population. Let allele frequencies in X be

$$p_X = 0.6,\ q_X = 0.4;\ r_X = 0.3,\ s_X = 0.7,$$

and in Y

$$p_Y = 0.2,\ q_Y = 0.8;\ r_Y = 0.5,\ s_Y = 0.5;$$

Table 10.2. F_1 genotype and gamete frequencies

Genotype	Frequency	*AB*	*Ab*	*aB*	*ab*
AB/AB	0.018	0.018			
AB/Ab	0.060	0.030	0.030		
AB/aB	0.084	0.042		0.042	
AB/ab	0.100	0.05–0.05c	0.05c	0.05c	0.05–0.05c
Ab/Ab	0.042		0.042		
Ab/aB	0.180	0.09c	0.09–0.09c	0.09–0.09c	0.09c
Ab/ab	0.196		0.098		0.098
aB/aB	0.048			0.048	
aB/ab	0.160			0.080	0.080
ab/ab	0.112				0.112
		$0.14+0.04c$	$0.26-0.04c$	$0.26-0.04c$	$0.34+0.04c$

and suppose each population to be in recombination equilibrium. The gamete frequencies g_{11}, g_{12}, g_{21}, g_{22} will be, respectively, 0.18, 0.42, 0.12, 0.28 in X, and 0.10, 0.10, 0.40, 0.40 in Y. Then the F_1 genotypic frequencies will be as shown in Table 10.2, producing gametes in the ratios also shown in Table 10.2.

In the cross population, p, q, r and s will each be the average of their values in X and Y, 0.4, 0.6, 0.4 and 0.6, respectively. Thus, equilibrium gamete frequencies in the population descending from the cross will be

$$g_{11e}: g_{12e}: g_{21e}: g_{22e} = 0.16: 0.24: 0.24: 0.36.$$

But if we let c equal 1/2, the F_1 gamete production is precisely equal to these equilibrium values. If we cross two populations which are in recombination equilibrium, the F_1 of the cross will produce gametes in equilibrium proportions if the two loci are not linked. Only if c is less than 1/2 will the cross produce disequilibrium that will persist.

Thus, crossing will not produce persistent recombination disequilibrium between two loci if they are not linked. Fisher (1930) showed that selection will tend to create and/or increase recombination disequilibrium. Disequilibrium thus created between nonlinked loci will not immediately disperse when selection ceases.

Lecture 10 Exercises

1. In Lecture 9 Exercise 1, we crossed mouse inbred strains BALB/c (*AAcc*) with C57BL/6 (*aaCC*). What would be the equilibrium distribution of gametes and of genotypes in the population descended from this cross?

2. In Lecture 9 Exercise 2, what is the equilibrium array of gametes for the population? Calculate the disequilibrium in generation t and in generation t+3, when c=1/2, and when c=1/4.

Lecture 11 Polyploidy

A normal human gamete, sperm or ovum, carries a set of 23 chromosomes, each member of the set differing from all others in size, centromere placement, banding pattern, and gene content. When sperm and ovum unite, the resulting zygote will have two identical sets (except for the XY pair in males), 23 pairs of chromosomes. The gamete, with only one set of chromosomes, is said to be haploid (from the Greek *haploos*; single); the zygote, with its two sets, diploid.

In some species, cells may contain three, four, or more sets of chromosomes. By extension of the above terms, a cell with three sets of chromosomes is said to be triploid, with four sets, tetraploid, etc. The general term polyploid is applied to any cell with more than two chromosome sets.

Polyploidy is rare in higher animals; indeed, the condition appears to be lethal in mammals, birds and reptiles. It can occur, however, in lower chordates and invertebrates, and is common in higher plants. Man has, in fact, deliberately induced polyploidy in many domestic plant species, because it often enhances luxuriance and vigor.

An allopolyploid is one in which there are complete sets of chromosomes from two or more species. Gametes from two different species may sometimes unite to form a viable hybrid (Stebbins 1950). The hybrid carries one haploid set of chromosomes from each species; meiosis is impossible, because the chromosomes cannot form pairs. The mule, offspring of a jackass and a mare (or its reciprocal, the henny) is an example of a sterile interspecific hybrid.

In plants, however, such a hybrid may occasionally produce offspring due to the duplication of each of the two haploid chromosome sets. The Russian geneticist Karpechenko (1927, 1928), for example, produced a viable hybrid between the cabbage, *Brassica oleracea*, and the radish, *Raphanus sativus*. Each parent species had 9 pairs of chromosomes; the hybrid had 18 chromosomes which did not pair at meiotic prophase. The hybrid was highly sterile, but did produce a few seeds which developed into apparently diploid plants with 18 chromosome pairs. Chromosomes from each species paired with their homologues from the same species, and the hybrid behaved genetically as if it were a normal diploid.

Such a hybrid is said to be amphidiploid; the new species formed by this hybridization, *Raphanobrassica*, is an example of an allopolyploid. A naturally occurring amphidiploid allopolyploid, the cord grass, *Spartina townsendii*, apparently arose from a hybrid between *S. stricta* and *S. alterniflora* (Huskins 1931). The hybrid has 63 chromosome pairs, the parent species 28 and 35, respectively (Mertens 1971).

An autopolyploid, on the other hand, carries more than two identical haploid sets of chromosomes from the same species. Therefore, a given chromosome can pair with any of its homologues. This difference in pairing causes differences between the population genetics of an autopolyploid and that of a diploid.

We will investigate the population genetics of the autotetraploid in some detail. Haldane (1930) published general formulae for any orthoploid (an autopolyploid with an even number of haploid chromosome sets, i.e., tetraploid, hexaploid, octoploid, etc.). We will assume that the four homologues of any one chromosome form two pairs which separate normally during meiosis, so that a gamete is diploid, carrying a random pair of complete haploid sets. The union of two gametes then restores the tetraploid condition in the next generation. We must also assume that there is no crossing over between the A locus and the centromere of the chromosome on which this locus occurs. Crow and Kimura (1970) give formulae applicable when such crossing over does occur.

Suppose, then, that in an autotetraploid the alleles *A* and *a* are segregating at an autosomal locus. Five tetraploid genotypes are possible, *AAAA*, *AAAa*, *AAaa*, *Aaaa* and *aaaa*, and there will be three types of diploid gametes, *AA*, *Aa* and *aa*. Let the frequency of *AA* gametes formed in generation t be x_t, of *Aa*, $2y_t$ and of *aa*, z_t, where $x_t+2y_t+z_t=1$. Allele frequencies are then

$$p_t=x_t+y_t,$$
$$q_t=y_t+z_t. \tag{11.1}$$

Random union among these gametes will reconstitute the five tetraploid genotypes in generation t+1, in the frequencies shown in Table 11.1. Note that the *AAaa* genotype can result from the union of a *AA* with a *aa* gamete, or from the union of two *Aa* gametes. Therefore, there are two terms in the frequency of this genotype.

Under the assumptions given above, if we were to number the four chromosomes carrying the A locus in a given individual 1 through 4, that individual would form equal numbers of gametes carrying each of the six possible pairs of chromosomes (1 with 2, 1 with 3, 1 with 4, 2 with 3, 2 with 4, and 3 with 4). If, then, the individual were an *AAaa* homozygote, it would form gametes in the ratio of one *AA* (chromosome 1 with 2): four *Aa* (1 with 3, 1 with 4, 2 with 3 and 2 with 4): one *aa* (3 with 4). Thus, when the tetraploids of generation t+1 form gametes, the proportionate gametic output is as shown in Table 11.1. The probability that a k-ploid parent with r *A* and k−r *a* alleles will form an m-ploid

Table 11.1. Gamete frequencies in an autotetraploid population

Genotype	Frequency	*AA*	*Aa*	*aa*
AAAA	x_t^2	1		
AAAa	$4x_ty_t$	1/2	1/2	
AAaa	$4y_t^2+2x_tz_t$	1/6	2/3	1/6
Aaaa	$4y_tz_t$		1/2	1/2
aaaa	z_t^2			

gamete ($m=k/2$) with s *A* and $m-s$ *a* alleles is (Haldane 1930):

$$\Pr(A^s a^{m-s} \text{ given } A^r a^{k-r})$$
$$=(m!)^2 r!(k-r)!/k!s!(m-s)!(m-r+s)!(r-s)! \qquad (11.2)$$

Multiplying the gamete frequencies by the tetraploid frequencies in Table 11.1, and summing each column, gives the overall frequencies for gametes formed in generation $t+1$:

$$x_{t+1}=x_t^2+2x_ty_t+2y_t^2/3+x_tz_t/3$$
$$=x_t^2+2x_ty_t+x_tz_t+2y_t^2/3-2x_tz_t/3$$
$$=x_t(x_t+2y_t+z_t)+2(y_t^2-x_tz_t)/3$$
$$=x_t+2(y_t^2-x_tz_t)/3;$$
$$2y_{t+1}=2y_t-4(y_t^2-x_tz_t)/3;$$
$$z_{t+1}=z_t+2(y_t^2-x_tz_t)/3. \qquad (11.3)$$

Allele frequencies remain constant;

$$p_{t+1}=x_{t+1}+y_{t+1}$$
$$=x_t+2(y_t^2-x_tz_t)/3+y_t-2(y_t^2-x_tz_t)/3$$
$$=x_t+y_t=p_t. \qquad (11.4)$$

A genes are exchanged between *AA* and *Aa* gametes, but are neither created nor destroyed.

The parallels between this situation and recombination between two loci are obvious. Again, the change in gamete frequency from generation t to generation $t+1$ is a function of the gamete frequencies in t,

$$K(y_t^2-x_tz_t),$$

where K is $+2/3$ for the change in x or z, $-4/3$ for the change in 2y.

Equilibrium again occurs when this function is zero. By analogy to the diploid situation, we might expect diploid gamete frequencies to be $x_e=p^2$, $2y_e=2pq$, and $z_e=q^2$ at equilibrium. These values indeed do make

$$y_e^2-x_ez_e=(pq)^2-p^2q^2=0. \qquad (11.5)$$

It follows that the equilibrium genotypic ratio is

AAAA: *AAAa*: *AAaa*: *Aaaa*: *aaaa*

$$=p^4: 4p^3q: 6p^2q^2: 4pq^3: q^4 \qquad (11.6)$$

The terms of this ratio are the terms of the expansion of $(p+q)^4$, just as the terms of the diploid equilibrium ratio are given by the terms of the expansion of $(p+q)^2$. In general, the expansion of $(p+q)^k$ will give the terms of the equilibrium genotypic ratio for a k-ploid with two alleles, whether the k-ploid population is a population of gametes or of adults. Furthermore, we have seen that expanding $(\Sigma p_i)^2$ gave the terms of the diploid equilibrium with n alleles. For a k-ploid with multiple alleles, these terms are given by expanding $(\Sigma p_i)^k$.

Equilibrium is not attained in a single generation. For arbitrary values of x_t, $2y_t$, z_t, the quadratic change function, $y_t^2 - x_t z_t$, is not zero. If we represent this change function by d_t, however,

$$d_{t+1} = (y_t - 2d_t/3)^2 - (x_t + 2d_t/3)(z_t + 2d_t/3)$$
$$= y_t^2 - x_t z_t - 4y_t d_t/3 + 4d_t^2/9 - 2x_t d_t/3 - 2z_t d_t/3 - 4d_t^2/9$$
$$= y_t^2 - x_t z_t - (2d_t/3)(2y_t + x_t + z_t)$$
$$= d_t - 2d_t/3 = d_t/3. \quad (11.7)$$

d is only 1/3 as large in generation t+1 as it was in generation t, and obviously goes to zero very rapidly. Haldane (1930) showed that for a hexaploid,

$$d_{t+1} = 0.4\, d_t;$$

the proportion of the generation t disequilibrium which remains in generation t+1 is greater for hexaploids than for tetraploids. As k increases, the rate at which d goes to zero decreases.

Lecture 11 Exercise

Suppose that we create an autotetraploid by causing cells to multiply genomes without cell division. If the original cell was heterozygous for a pair of alleles, *A* and *a*, the tetraploid developed from that cell would be *AAaa*. What will be the gamete frequencies in the first three generations? What will the equilibrium genotypic ratio be?

Part III Systematic Forces

Lecture 12 Selection

The assumed conditions under which we derived the Hardy-Weinberg law precluded any change in allele frequencies, because we assumed that none of the forces that could produce such changes existed. We have considered the consequences of several departures from the original assumptions, but these have not yet permitted the introduction of such forces. Even when we crossed populations that differed in allele frequency, a new frequency was immediately established in the cross population which was constant thereafter. Therefore, allele frequencies have been constant. Constant allele frequencies made possible the establishment of a genotypic equilibrium in each situation.

We must now begin to consider the effect of introducing changes in allele frequencies. Falconer (1960, 1981) recognized two general types of forces, systematic and dispersive, which can cause such changes. Systematic forces cause changes that are predictable in both magnitude and direction. Changes caused by dispersive forces are predictable in magnitude, but not in direction. We will first investigate the systematic forces, beginning with selection.

Soon after the rediscovery of Mendel's laws, Cuenot (1905, cited by Castle and Little 1910) found that he was unable to produce homozygous yellow mice. All his matings between two yellow mice always produced both yellow and black offspring, indicating that his yellow mice were heterozygous $A^y a$; black mice were known to be homozygous *aa*. Cuenot suggested that A^y ova could not, for some reason, be fertilized by A^y sperm. Castle and Little (1910) found a 2: 1 ratio of $A^y a$ to *aa* from yellow x yellow matings, and suggested that $A^y A^y$ zygotes were formed in more or less normal frequency, but were unable to develop. Ibsen and Steigleder (1917) confirmed this suggestion by showing that 1/4 of implanted embryos from $A^y a \times A^y a$ matings, presumably the $A^y A^y$ homozygotes, died *in utero*. The $A^y A^y$ genotype in the mouse is an example of an obligate lethal genotype. The A^y gene is dominant with respect to coat color, heterozygotes being yellow, but recessive with respect to its lethal effect.

Lethal genotypes have been identified in all species whose genetics have been investigated extensively. Castle and Little, as early as 1910, refer to lethal chlorophylless mutants in plants. Ibsen and Steigleder (1917) cite lethals as being well known in *Drosophila*. Lethal effects tend to be recessive, since a gene with completely dominant lethal effects could not be transmitted.

Let us suppose, then, that we have a diploid population in which a pair of alleles, *M* and *m*, are segregating at an autosomal locus. If the frequencies of the two alleles are p_t and q_t, respectively, among gametes produced by generation t, and these gametes unite at random to form the zygotes of generation $t+1$, the genotypic ratio among those zygotes will be

$$MM: Mm: mm = p_t^2: 2p_tq_t: q_t^2.$$

Suppose further that the *m* allele has recessive lethal effects; *mm* zygotes fail to develop to term, or die at some time before reaching sexual maturity. The ratio of genotypes among individuals producing gametes in generation t+1 will be

$$MM: Mm: mm = p_t^2: 2p_tq_t: 0. \tag{12.1}$$

The *mm* genotypes, being no longer alive, cannot produce gametes.

The total number of gametes formed will then be proportional to $p_t^2 + 2p_tq_t$; p_t of these will carry *M* alleles, p_tq_t (half the frequency of heterozygotes) will carry *m*. Thus, allele frequencies among the gametes formed by t+1 will be

$$\begin{aligned} p_{t+1} &= p_t/(p_t^2 + 2p_tq_t) = p_t/(p_t)(p_t + 2q_t) \\ &= 1/(1+q_t); \\ q_{t+1} &= p_tq_t/(p_t^2 + 2p_tq_t) = q_t/(1+q_t). \end{aligned} \tag{12.2}$$

(Check: these frequencies do sum to 1.)

The original frequency of *m*, q_t, can, of course, be written as $q_t/1$. Therefore, $q_{t+1} < q_t$; both have the same numerator, but q_{t+1} has the larger denominator. Lethality of *mm* homozygotes leads, not surprisingly, to a decrease in *m* frequency.

Because *mm* is equally lethal in every generation, the relationship between q_{t+1} and q_t expressed in Eq. (12.2) is recursive, applying to any two successive generations:

$$\begin{aligned} q_{t+2} &= q_{t+1}/(1+q_{t+1}) = [q_t/(1+q_t)]/\{1 + [q_t/(1+q_t)]\} \\ &= [q_t/(1+q_t)]/[(1+q_t+q_t)/(1+q_t)] \\ &= q_t/(1+2q_t); \\ q_{t+3} &= [q_t/(1+2q_t)]/[(1+2q_t+q_t)/(1+2q_t)] \\ &= q_t/(1+3q_t); \\ q_{t+n} &= q_t/(1+nq_t). \end{aligned} \tag{12.3}$$

The numerator of this fraction is constant; the denominator increases continuously as n increases, so that q_{t+n} will go to zero as n becomes very large. Eventually, lethality of the *mm* genotype will lead to elimination of the *m* allele.

We conventionally express changes in the frequency of the *m* allele in terms of Δq_t, the change in q between generations t and t+1:

$$\Delta q_t = q_{t+1} - q_t. \tag{12.4}$$

Expressed in this form, Δq_t is negative if q decreases, positive if q increases, between t and t+1. Because $p+q=1$ in every generation, of course, any change in q is necessarily accompanied by an equal change in the opposite direction in p; $\Delta p_t = -\Delta q_t$.

In the specific case where *mm* is an obligate lethal genotype,

$$\Delta q_t = [q_t/(1+q_t)] - q_t = (q_t - q_t - q_t^2)/(1+q_t)$$
$$= -q_t^2/(1+q_t) \qquad (12.5)$$

The *mm* genotype, present among zygotes of generation t+1 can not contribute to the pool from which zygotes of generation t+2 will be formed. As a result the frequency of *m* alleles will be less among generation t+1 gametes than it was among those from t. This constitutes selection against the *mm* genotype.

We will define selection as any situation in which a genotype cannot contribute to the gamete pool in proportion to its frequency among zygotes, due to reduced viability and/or fertility. Like Wright (1955), we will exclude from selection disproportionate contributions of zygotic genotypes to the gamete pool arising from sampling error, from migration, or from mutation.

If a lethal genotype expresses itself only after reproduction takes place, there will be no disproportionate contribution to the gametes, and therefore no selection. The death of *mm* homozygotes after they have reproduced obviously does not prevent them from contributing to the gamete pool. Allelic frequencies among gametes produced by generation t+1 will be p_t, q_t; zygotic genotypic frequencies will be the same in generation t+2 as they were in t+1. Huntington's disease, a fatal neural disorder in man, is caused by a dominant gene (Thompson and Thompson 1980). The average age of onset of the symptoms of this disease is 38 years; many who will later develop the disease will have reproduced before they know that they have it. Thompson and Thompson, in fact, do not classify the Huntington's disease gene as a lethal, defining lethal genes as those that cause death before sexual maturity (a curious confusion, it would seem, between lethal effect and selection effect.)

As our definition of selection implied, a genotype need not be lethal to cause selection. Suppose that the *mm* homozygotes are perfectly viable, but totally sterile. Perhaps they produce no gametes at all, or perhaps they produce gametes which are not functional. In either case, though *mm* individuals are alive when their generation is reproducing, they are no more capable of participating in that reproduction than if they were dead. The effective ratio of genotypes at reproduction is that given by Eq. (12.1), and allele frequencies would change as in Eq. (12.5).

Lethality and/or sterility of a genotype need not be complete to cause selection. Suppose that only a fraction, s, of *mm* homozygotes die or are sterile. The genotype's participation in reproduction would still be reduced. With only 1−s of the *mm*'s reproducing, Eqs. (12.1), (12.2), and (12.5) can be written as

$$MM\colon Mm\colon mm = p_t^2\colon 2p_tq_t\colon (1-s)q_t^2; \qquad (12.1')$$

$$q_{t+1} = [p_tq_t + (1-s)q_t^2]/[p_t^2 + 2p_tq_t + (1-s)q_t^2]$$
$$= (p_tq_t + q_t^2 - sq_t^2)/(p_t^2 + 2p_tq_t + q_t^2 - sq_t^2)$$
$$= (q_t - sq_t^2)/(1 - sq_t^2); \qquad (12.2')$$

$$\Delta q_t = [(q_t - sq_t^2)/(1 - sq_t^2)] - q_t$$
$$= (q_t - sq_t^2 - q_t + sq_t^3)/(1 - sq_t^2)$$
$$= -sq_t^2(1 - q_t)/(1 - sq_t^2) = -sp_tq_t^2/(1 - sq_t^2). \qquad (12.5')$$

q will decrease, though less rapidly than when all *mm* died or were sterile. If we substitute $s = 1$ into Eq. (12.5′),

$$\Delta q_t = -p_tq_t^2/(1 - q_t^2) = -(1 - q_t)q_t^2/(1 - q_t)(1 + q_t)$$
$$= -q_t^2/(1 + q_t);$$

Eq. (12.5′) reduces to Eq. (12.5). Complete lethality or sterility is merely a special case of selection which arises when $s = 1$. Equations (12.1′) and (12.2′) likewise reduce to (12.1) and (12.2) when $s = 1$.

At the other extreme, if $s = 0$, Δq_t is 0, from Eq. (12.5′). There is no change in allele frequency; the participation of *mm* is not decreased, Eq. (12.1′) reducing to the Hardy-Weinberg ratio; and Eq. (12.2′) reduces to $q_{t+1} = q_t$. There is no selection.

Even if all *mm* homozygotes survive and are capable of reproduction, their fecundity may be reduced. That is to say, they may produce fewer offspring than other genotypes. In mice, for example, *mm* sires and dams might produce smaller litters than sires and dams of other genotypes. The effective frequency of *mm* genotypes during reproduction is reduced by a fraction s representing the reduction in litter size; thus, there is selection against *mm*.

The fraction s represents the overall reduction in the effective frequency of *mm* homozygotes in the reproducing population due to all causes of reduction. Suppose, for example, that half the *mm* die; that a fourth of the survivors are sterile; and that the fecundity of the fertile survivors is reduced by 20%. The effective frequency of this genotype at reproduction is then

$$(1 - 0.5)(1 - 0.25)(1 - 0.2)q_t^2 = 0.3q_t^2.$$

As this represents $(1 - s)q_t^2$, $s = 0.7$.

The fraction s, which measures the reduction in the participation of *mm* in reproduction, is called the selection coefficient for genotype *mm*. It represents the selective disadvantage of *mm*, the extent to which that genotype's contribution to reproduction is reduced.

Both natural and artificial selection operate through reduced participation of a genotype in reproduction. Lethality of A^yA^y is a natural phenomenon, and selection against A^yA^y is natural selection. Suppose, however, that *mm* individuals are normally viable and fertile; but in a domesticated population in which they arise, the breeder wishes to reduce the frequency of the *m* gene. He can kill the *mm* homozygotes before they reproduce, introducing an artificial lethality. Or he can keep them alive, but introduce an artificial sterility by castrating them, or merely by excluding them from reproduction. These actions of the breeder are equivalent, in their effect on *m* allele frequency, to naturally occurring lethality or sterility. The breeder has imposed artificial selection on the population; but the means by which this selection takes effect are the same as those by which natural selection would take effect.

In any population under artificial selection, both artificial and natural selection will be occurring simultaneously. Often nature and the breeder will be selecting in the same direction. The breeder must, after all, produce an organism capable of surviving and reproducing, regardless of what other qualities he may wish to incorporate. Therefore, both natural selection and the breeder will tend to select against lethal or sterile genotypes. In such cases, the selection coefficient is the result of both natural and artificial selection. Suppose, for example, that half the *mm* die in utero. The breeder, aware of this, may want to reduce the frequency of *m* by excluding the surviving *mm* from his choice of mates, even though these individuals are normally fertile. Then no *mm* will be allowed to reproduce; although the natural selection coefficient, s_n, is only 0.5, the artificial selection coefficient, s_a, is 1 for the survivors, and

$$(1-s_n)(1-s_a)=(0.5)(0)=0.$$

The above example serves to illustrate an important difference between natural and artificial selection. Although completely lethal genotypes do occur in nature, natural selection coefficients are more frequently relatively small. This is probably because a large s_n tends to eliminate the lethal gene very quickly; most alleles that remain in a population for any length of time are likely to be at only a slight disadvantage. The breeder, on the other hand, is likely to impose an $s_a=1$. If he wishes to eliminate *m* from the population, it would be foolish to use any *mm* for breeders unless absolutely necessary.

Often, however, the breeder may wish to increase the frequency of a genotype that is at a natural selection disadvantage. The *mm* genotype may be esthetically pleasing, or may have other advantages in the eyes of the breeder; or he may merely wish to preserve and increase the deleterious genotype for study. He must, then, attempt to counter the effects of natural selection.

It may be possible to do this by changing the environment. The sex-linked gene naked (*n*) in domestic fowl causes approximately 50% mortality prior to hatching in *nn* males and *n* females (Hutt 1949). Survivors are featherless, and under normal brooding conditions, another 50% of each sex will die in the brooder. But if brooder temperatures are elevated, post-hatching mortality can be greatly decreased. This is an example of a facultative lethal, as opposed to obligate lethals such as A^yA^y.

More frequently, however, s_n cannot be decreased in any practicable environment. The only way to maintain the deleterious gene in the population is to impose countervailing selection against the other allele. A^y can be maintained in a mouse population only by restricting the breeding of *aa* homozygotes; only $A^ya \times A^ya$, or occasionally $A^ya \times aa$, matings are used.

Under these circumstances, there will be selection against two genotypes, natural selection against *mm* and artificial selection against *MM*. In many cases, also, if *mm* is deleterious, the heterozygote *Mm* may also be at a natural selective disadvantage. The breeder trying to eliminate *m* alleles from his population may also try to restrict the breeding of *Mm* as well as that of *mm*. Our discussions so far have centered on selection against only one genotype. The effects of selection against more than one genotype at a time are best described in terms of relative fitnesses and the general selection equation.

Lecture 12 Exercises

1. A certain autosomal recessive mutation (*m*) in mice is lethal to about 1/3 of the *mm* homozygotes of either sex before the pups are weaned. Among surviving homozygotes, half are sterile, and the remainder produce 8 offspring per litter, on the average, compared to 10 offspring per litter from *MM* and *Mm* parents. Heterozygotes are completely normal with respect to viability and fertility. Determine the distribution of genotypes forming gametes in generation $t+1$, and calculate q_{t+1}, in terms of q_t.

2. Calculate Δq_t for Exercise 1, when $q_t = 0.25$, and when $q_t = 0.4$. What will q_{t+3} be in each case?

Lecture 13 The General Selection Equation

In his monumental *Origin of Species* (1859), Charles Darwin entitled the fourth chapter *Natural Selection, or the Survival of the Fittest.* Thus he introduced into biological science the concept of fitness, linked with the concept of selection.

What Darwin meant by fitness was simply the ability to produce offspring. Quantitatively, we can measure the fitness of a given genotype as the proportion of fertile individuals of that genotype which survives to reproduce. The fitness of genotype *MM* is greater than that of *mm* if a larger proportion of fertile *MM* individuals than of *mm* survives.

If all genotypes are equally fit, no selection will take place. Suppose that among 100 zygotes in generation $t+1$ of a population, 36 are *MM*, 48 *Mm* and 16 *mm*. The population lives in a very harsh environment, so that only 50 of the 100 zygotes will survive. If those 50 survivors comprise 18 *MM*, 24 *Mm* and 8 *mm*, the proportion of survivors is 0.5 for each genotype. Allele frequencies among survivors are $p = 0.6$, $q = 0.4$, as they were among the original zygotes. Assuming equal fertility of all survivors, $\Delta q_t = 0$; there is no selection. Only if the proportion of fertile survivors differs among the genotypes will selection take place.

The relative survival of the genotypes is thus more important than the absolute survival proportions in determining selection. It is convenient to express relative survival proportions by dividing the observed proportions by the largest of them. For example, if the 50 survivors of the population above had comprised 30 *MM*, 18 *Mm* and 2 *mm*, survival proportions would have been 0.833, 0.375 and 0.125, respectively. Dividing each of these by the largest, we obtain

$$w_{MM} = 0.833/0.833 = 1;$$

$$w_{Mm} = 0.375/0.833 = 0.45;$$

$$w_{mm} = 0.125/0.833 = 0.15.$$

These w values are referred to as the relative fitnesses of the three genotypes. The relative fitness of the genotype with the largest survival proportion will always be 1, with the fitnesses of the other genotypes defined relative to that value.

We will assign the symbols w_{11} and w_{22} to the fitnesses of the two homozygotes, while the fitness of the heterozygote will be w_{12}. Usually, w_{11} will be assigned to the homozygote with the higher fitness value. Thus, if *m* is a deleterious gene, w_{11} will be the fitness of the *MM* homozygote.

It is, of course, possible for two genotypes to have the same fitness value, if they have equal survival proportions. If 30 *MM* and 40 *Mm* had survived in the above population, both would have had survival fractions of 0.833, and

$$w_{11} = w_{12} = 1.$$

We have been discussing fitness in terms of survival proportions, assuming equal fertility of all survivors. In fact, fertility differences compound with differences in survival to determine relative fitness. If all 16 *mm* individuals had survived, but only 2 were fertile, w_{mm} would still be 0.15.

The fitness value is obviously related to the selection coefficient; w is the relative proportion of fertile survivors, s the proportion lost through death or infertility. Therefore,

$$w = 1 - s \tag{13.1}$$

for any genotype. In describing selection against the *mm* genotype only we tacitly assumed

$$w_{11} = w_{12} = 1,$$

implying

$$s_{11} = s_{12} = 0.$$

We can calculate the average fitness in generation t, $\bar{W}_t$, as the average of the fitness values weighted by the zygotic frequencies of the genotypes:

$$\bar{W}_t = w_{11}p_t^2 + 2w_{12}p_tq_t + w_{22}q_t^2.$$

If we suppose that $w_{11} = 1$, and substitute in terms of selection coefficients for the others,

$$\begin{aligned}\bar{W}_t &= p_t^2 + 2(1 - s_{12})p_tq_t + (1 - s_{22})q_t^2 \\ &= 1 - 2s_{12}p_tq_t - s_{22}q_t^2.\end{aligned} \tag{13.2}$$

The heterozygote for a deleterious gene often shows abnormalities less severe in effect than those shown by the homozygote. In mice, for example, the varitint-waddler (*Va*) gene described by Cloudman and Bunker (1945) is listed by Green (1966) as semidominant. Heterozygous *Vava* mice differ from "wild-type" mice in having a variegated coat color pattern, abnormal behavior and degeneration of the middle ear. Homozygous *VaVa* mice show the same abnormalities in more severe form.

If fitness also exhibits partial dominance, the reduction in fitness of *Mm* will be smaller than the reduction for *mm*. If we calculate the fraction $h = s_{12}/s_{22}$, we can let $s_{22} = s$, $s_{12} = hs$. The average fitness can then be written as

$$\bar{W}_t = 1 - 2hsp_tq_t - sq_t^2. \tag{13.3}$$

Selection changes p and q, and therefore changes $\bar{W}$ even if the w's are constant. In fact, $\bar{W}$ will increase as the frequencies of the less fit genotypes decrease under selection. It should therefore be possible to relate Δq to $\bar{W}$. This relationship is expressed by the general selection equation,

$$\Delta q_t = (p_tq_t/k\bar{W}_t)(d\bar{W}_t/dq_t), \tag{13.4}$$

first stated by Wright (1937). k in this equation is the ploidy level, 2 for most of the applications we will make. Li (1967b) derived the more general form

$$\Delta q_t = (p_tq_t/k\bar{W}_t)(b_{wx}),$$

where b_{wx} is the regression of fitness on a variable x which takes on the values 0, 1/2 and 1 for genotypes *MM*, *Mm* and *mm*, respectively. Li's equation reduces to Eq. (13.4) under random mating.

We can apply the general selection equation to the situation with partial dominance for fitness, as defined in Eq. (13.3). Substituting $p_t = 1 - q_t$ in Eq. (13.3), we define $\bar{W}_t$ as a series of terms in q_t, s and h:

$$\bar{W}_t = 1 - 2hsq_t + 2hsq_t^2 - sq_t^2.$$

The derivative of $\bar{W}_t$ with respect to q_t is then the sum of the derivatives of the terms in the above equation, term by term:

$$d\bar{W}_t/dq_t = d(1)/dq_t + d(-2hsq_t)/q_t + d(2hsq_t^2)/dq_t + d(-sq_t^2)/dq_t. \quad (13.5)$$

We can evaluate these derivatives easily, because

$$d(ax^v)/dx = vax^{v-1}.$$

Therefore, Eq. (13.5) can be written as

$$d\bar{W}_t/dq_t = 0(1)q_t^{-1} + 1(-2hsq_t^0) + 2(2hsq_t^1) + 2(-sq_t^1)$$
$$= 0 - 2hs + 4hsq_t - 2sq_t.$$

Substituting this result into Eq. (13.4),

$$\Delta q_t = (p_t q_t / 2\bar{W}_t)(-2hs + 4hsq_t - 2sq_t)$$
$$= (-sp_t q_t)(h - 2hq_t + q_t)/(1 - 2hsp_t q_t - sq_t^2). \quad (13.6)$$

Equation (13.6) represents the general solution where *MM* has maximum fitness; we can refer to this as the superior homozygote case. There is selection against both *Mm* and *mm*, but the selection coefficient for *Mm* is a fraction, h ($0 \leqq h \leqq 1$), of that for *mm*.

If there is no selection against the heterozygote, $h = 0$. This is equivalent to saying that the fitness effects of *m* are recessive. If we substitute $h = 0$ into Eq. (13.6),

$$\Delta q_t = -sp_t q_t(0 - 0 + q_t)/(1 - 0 - sq_t^2)$$
$$= -sp_t q_t^2/(1 - sq_t^2). \quad (13.7a)$$

This is Eq. (12.5′), where we assumed selection against *mm* only.

If, on the other hand, $h = 1$, the fitness effects of *m* are completely dominant; selection against the heterozygote is as strong as against the homozygote. Then

$$q_t = -sp_t q_t\,(1 - 2q_t + q_t)/(1 - 2sq_t + 2sq_t^2 - sq_t^2)$$
$$= -sp_t^2 q_t/(1 - 2sq_t + sq_t^2). \quad (13.7b)$$

In this situation, if $s = 1$,

$$\Delta q_t = -p_t^2 q_t/(1 - 2q_t + q_t^2) = -p_t^2 q_t/p_t^2$$
$$= -q_t,$$

and the frequency of *m* in generation t+1 will be

$$q_{t+1}=q_t+\Delta q_t=q_t-q_t=0.$$

If no individual carrying the *m* gene reproduces, either homozygote or heterozygote, that allele will be eliminated in a single generation.

In many cases, of course, h will lie between 0 and 1. The fitness effects of *m* are partially dominant. As an example, suppose heterozygotes suffer exactly half the reduction in fitness of the homozygote; h=1/2. This is the so-called "no dominance" situation. Then

$$\begin{aligned}\Delta q_t &= -sp_tq_t[(1/2)-q_t+q_t]/(1-sq_t+sq_t^2-sq_t^2)\\ &= -sp_tq_t/2(1-sq_t).\end{aligned} \tag{13.7c}$$

Δq_t is a function, not only of s and h, but of p_t, q_t, and of the product p_tq_t. The change in allele frequency will be at a maximum when $p_t=q_t=0.5$, because p_tq_t is then maximized. As p_t increases beyond 0.5 due to selection, the rate of increase will slow. The eventual result of selection is indeed the elimination of the deleterious gene, but that result is approached asymptotically, the rate of approach slowing as q becomes very small. For example, with s=0.5 and h=0,

$$\Delta q_t=-0.07 \text{ when } p_t=q_t=0.5,$$

$$\Delta q_t=-0.004 \text{ when } p_t=0.9,\ q_t=0.1.$$

If we assume s=0.5, h=1,

$$\Delta q_t=-0.1 \text{ when } p_t=0.5,$$

$$\Delta q_t=-0.04 \text{ when } p_t=0.9.$$

Selection can also take place among gametes, if gametes carrying different genes vary in viability and/or fertility. Suppose that the viability or fertility of *m* gametes formed by generation t parents is reduced by a fraction s*, relative to that of *M* gametes. Then the frequency of *m* in generation t+1 will be

$$q_{t+1}=(1-s^*)q_t/[p_t+(1-s^*)q_t]=(1-s^*)q_t/(1-s^*q_t),$$

so that

$$\begin{aligned}\Delta q_t &= (q_t-s^*q_t-q_t+s^*q_t^2)/(1-s^*q_t)\\ &= -s^*q_t(1-q_t)/(1-s^*q_t)=-s^*p_tq_t/(1-s^*q_t).\end{aligned} \tag{13.8}$$

Alternatively, we can apply the general selection equation to this case of gamete selection. Among haploid gametes we have only two genotypes, *M* and *m*. These have fitnesses $w_1=1$ and $w_2=1-s^*$, respectively, so that

$$\bar{W}_t=w_1p_t+w_2q_t=p_t+(1-s^*)q_t=1-s^*q_t;$$

$$d\bar{W}_t/dq_t=-s^*.$$

Remembering that since gametes are haploid, k=1, Eq. (13.4) gives us

$$\begin{aligned}\Delta q_t &= (p_tq_t/\bar{W})(-s^*)\\ &= -s^*p_tq_t/(1-s^*q_t).\end{aligned}$$

This is the same result as we obtained by direct calculation in Eq. (13.8). The general selection equation can be applied to haploids, and indeed to any level of ploidy, with appropriate values of k. Li (1967b) discusses the application of the general selection equation to different levels of ploidy and to multiple alleles, reviewing a considerable literature.

Gamete selection, as described above, acts against all gametes produced by *mm* and half of those produced by *Mm* genotypes. It might therefore superficially appear to resemble superior homozygote selection with $h = 1/2$. But Eq. (13.8) is not equivalent to Eq. (13.7c). Δq_t is only half as large under superior homozygote selection with $h = 1/2$ as under gamete selection. The reason is easy to see. Gamete selection against *m* reduces the fitness of *m* gametes only. Selection against the *Mm* genotype reduces the fitness of both *M* and *m* gametes, so that the relative disadvantage of *m* is less under superior homozygote selection with $h = 1/2$ than under gamete selection.

Lecture 13 Exercise

In a population of a certain animal species, we decide to eliminate the gene *a* by selection. The frequency of *a* is $q_t = 0.5$, at present, and no selection has yet taken place. We can make a maximum of 75 single pair matings per generation in this species, and we want to keep population size (number of matings) as large as we can. From our 75 single pair matings, we can expect to produce only about 100 offspring of each sex; any kind of polygamous mating will yield even fewer offspring. The three genotypes *AA*, *Aa* and *aa* are phenotypically distinguishable, and we can cull either *Aa* or *aa* individuals, or both, to eliminate *a* genes; that is, we can set values of both s and h. What combination of s and h will give the maximum negative Δq_t in generation t, and what will this Δq_t be? How should we proceed in future generations?

Lecture 14 Equilibrium under Selection

Selection, as we have defined it, changes allele frequencies, resulting in corresponding changes in frequencies of genotypes. Equilibrium cannot exist if allele frequencies are changing, because allele frequency changes change the frequencies of genotypes.

Δq_t represents the change in allele frequency per generation. For equilibrium to exist, allele frequencies must be constant; only if

$$\Delta q_e = (p_e q_e / 2\bar{W}_e)(d\bar{W}_e/dq_e) = 0, \tag{14.1}$$

can equilibrium occur.

$\bar{W}_e$ is a function of genotypic frequencies and relative fitness values. Its maximum value, from Eq. (13.4), is 1. Therefore, Eq. (14.1) cannot be driven to 0 because its denominator, $2\bar{W}_e$, becomes infinitely large.

If either p or q is zero, Eq. (14.1) will be zero. With only one allele in the population, selection cannot change allele frequencies. In fact, selection as we have defined it cannot even take place, since with only one allele, there can be only one genotype, and we cannot have differential fitnesses of different genotypes. "Equilibrium" points with p or q equal to zero are therefore trivial.

Equation (14.1) can also be zero if the derivative of $\bar{W}_e$ is zero. In its most general terms, we can write $\bar{W}_t$ as

$$\begin{aligned}\bar{W}_t &= w_{11}(1-q_t)^2 + 2w_{12}(1-q_t)q_t + w_{22}q_t^2 \\ &= w_{11} - 2w_{11}q_t + w_{11}q_t^2 + 2w_{12}q_t - 2w_{12}q_t^2 + w_{22}q_t^2.\end{aligned}$$

The derivative is then

$$\begin{aligned}d\bar{W}_t/dq_t &= -2w_{11} + 2w_{11}q_t + 2w_{12} - 4w_{12}q_t + 2w_{22}q_t \\ &= -2(1-q_t)w_{11} + 2(1-q_t)w_{12} - 2w_{12}q_t + 2w_{22}q_t \\ &= 2p_t(w_{12} - w_{11}) - 2q_t(w_{12} - w_{22}).\end{aligned}$$

It follows that if we set the derivative equal to zero at equilibrium,

$$p_e(w_{12} - w_{11}) = q_e(w_{12} - w_{22}). \tag{14.2}$$

If the difference $w_{12} - w_{11}$ and the difference $w_{12} - w_{22}$ have the same sign, Δq_e can be 0 even if neither p nor q is 0. But for these two differences to have the same sign, the heterozygote must be more fit or less fit than either homozygote.

This situation is referred to as overdominance; positive overdominance if the heterozygote is more fit than either homozygote, negative overdominance if it is less. Some authors reserve the term overdominance for positive overdominance, referring to negative overdominance as "underdominance". The positive-negative dichotomy appears preferable, and this is the terminology we shall use.

The classic example of positive overdominance was elucidated by Allison (1954). The sickle cell anemia gene produces abnormal hemoglobin; homozygotes for this allele suffer from severe anemia, with the further complication that the erythrocytes deform into a sickle shape which can clog minor blood vessels. These homozygotes usually die by their early teens. About 40% of the hemoglobin of heterozygotes is abnormal sickling hemoglobin, but they are clinically normal. The presence in these heterozygotes of the sickling gene and the abnormal hemoglobin which it produces can be detected by appropriate blood tests; for example, subjecting heterozygotes' blood to abnormally low oxygen tensions will cause sickling of some of the erythrocytes.

The severe effects of the sickling homozygous condition would lead us to expect it to be in very low frequency in human populations. But Allison found this allele in very high frequencies, approaching 0.5, in some African populations living in malarial regions. Further study showed that the heterozygote was less susceptible to malaria than homozygous normals. Thus, in malarial regions, sickle cell homozygotes are selected against because of their abnormal hemoglobin; but there is countervailing selection against normal homozygotes due to the endemic malaria. Only heterozygotes can breed normally. The heterozygote thus has superior fitness due to the selective disadvantages of each of the two homozygotes.

With positive overdominance, the ratio of the fitness values is

$$w_{11}\colon w_{12}\colon w_{22} = 1-s_1\colon 1\colon 1-s_2,$$

so that at equilibrium

$$\begin{aligned} &p_e(1-1+s_1)=q_e(1-1+s_2),\\ &s_1(1-q_e)=s_2q_e,\\ &q_e(s_1+s_2)=s_1,\\ &q_e=s_1/(s_1+s_2);\ p_e=s_2/(s_1+s_2). \end{aligned} \tag{14.3}$$

Equation (14.3) says that the equilibrium frequency of either allele is positively correlated with the selection coefficient against the opposite homozygote. This is intuitively sensible. If selection against *MM* increases, the frequency of *M* should decrease, and that of *m* increase. If there is equal selection against both homozygotes, p_e and q_e should be equal.

With positive overdominance,

$$\begin{aligned} \bar{W}_t &= 1-s_1p_t^2-s_2q_t^2\\ &= 1-s_1+2s_1q_t-s_1q_t^2-s_2q_t^2, \end{aligned}$$

and therefore

$$\begin{aligned} d\bar{W}_t/dq_t &= 2s_1-2s_1q_t-2s_2q_t\\ &= 2[s_1-(s_1+s_2)q_t]. \end{aligned} \tag{14.4}$$

Obviously, if $q_t=q_e=s_1/(s_1+s_2)$ in Eq. (14.4),

$$d\bar{W}_t/dq_t=0.$$

Table 14.1. Positive overdominance; approach to equilibrium

Generation	q_t	$q_e - q_t$	$\bar{W}_t$	$d\bar{W}_t/dq_t$	$p_t q_t$	Δq_t
0	0.5	−0.1	0.75	−0.2	0.25	−0.0333
1	0.4667	−0.0667	0.7555	−0.1334	0.2489	−0.0220
2	0.4447	−0.0447	0.7580	−0.0894	0.2469	−0.0146
0	0.3	0.1	0.75	0.2	0.21	0.028
1	0.328	0.072	0.7548	0.144	0.2204	0.0210
2	0.349	0.051	0.7574	0.102	0.2272	0.0153

Now consider what happens when $q_t \neq q_e$. Suppose that

$$p_t = p_e + x; \; q_t = q_e - x.$$

Then from Eq. (14.4),

$$d\bar{W}_t/dq_t = 2s_1 - 2(s_1 + s_2)q_e + 2(s_1 + s_2)x$$
$$= 2(s_1 + s_2)x.$$

Since $2(s_1 + s_2)$ is always positive, the sign of the partial derivative will be the same as the sign of x. And since $p_t q_t / 2\bar{W}_t$ is also always positive, the sign of Δq_t will also always be the same as the sign of x.

If, then, x is positive, q will be increasing. But x will be positive if $q_t < q_e$. x will be negative, and q will decrease, only if $q_t > q_e$. In either case, q_{t+1} will be closer to q_e than was q_t. q approaches its equilibrium value whether initially reater or initially less than that value.

Table 14.1 illustrates the approach to equilibrium under positive overdominance, assuming specific values $s_1 = 0.4$, $s_2 = 0.6$, $q_e = 0.4$. Two initial q values are illustrated. It is not necessary, of course, that $s_1 + s_2 = 1$, but it is convenient for the example to make them so. This does not restrict the generality of the result.

Note that Δq_t decreases as q approaches q_e. The approach to equilibrium is again asymptotic, the change in q becoming smaller as x decreases.

If the heterozygote is inferior in fitness to both homozygotes, we have negative overdominance. There is no outstanding example of negative overdominance in the literature, comparable to the sickle cell anemia case as an example of positive overdominance. One possible cause of this may be the ephemeral nature of negative overdominance equilibria, as we shall see.

With negative overdominance, the ratio of the fitnesses can be written

$$w_{11} : w_{12} : w_{22} = 1 : 1 - s : 1 - h's.$$

h′, like h, is a fraction between 0 and 1. Although both *Mm* and *mm* are selected against, selection against *Mm* is more severe. If $h' = 0$, there is no selection against either homozygote; both have fitnesses 1. If $h' = 1$, of course, we no longer have negative overdominance, but selection with a superior homozygote, as in Eq. (13.8b).

With the above negative overdominance fitness ratio,

$$\bar{W}_t = 1 - 2sp_t q_t - h'sq_t^2;$$

$$d\bar{W}_t/dq_t = -2s + 4sq_t - 2h'sq_t = -2s[1 - (2 - h')q_t]. \quad (14.5)$$

Setting this equation equal to zero, we can solve for the equilibrium frequency of *m*, q_e:

$$2s(2-h')q_e - 2s = 0;$$

$$q_e = 1/(2-h'). \quad (14.7)$$

It is rather surprising that the equilibrium frequency of *m* is independent of s, the strength of selection against the heterozygote. It depends only on the ratio of the homozygote disadvantage relative to that of the heterozygote. If all the *Mm* heterozygotes and half the *mm* homozygotes failed to reproduce, we might expect a lower q_e than if half the heterozygotes and a quarter of the homozygotes failed. Yet in both cases, $h' = 0.5$ and q_e is therefore 0.6667.

When $h' = 0$, there is no selection against either homozygote, as noted above. Then $q_e = 0.5$, which is intuitively reasonable. As we introduce increasing selection against *mm* by increasing h', we might expect q_e to decrease. In fact, q_e increases as h' increases; we have just noted the example of $q_e = 0.6667$ when $h' = 0.5$. This again is a surprising result.

Let us look at the stability of the equilibrium under negative overdominance. Again we let

$$p_t = p_e + x; \; q_t = q_e - x.$$

The derivative is then

$$d\bar{W}_t/dq_t = -2s + 2s(2-h')q_e - 2s(2-h')x$$
$$= -2s(2-h')x.$$

$2s(2-h')$ is alwas positive; therefore, $d\bar{W}_t/dq_t$ and Δq_t will have the opposite sign to that of x. x is positive if $q_t < q_e$; but then Δq_t is negative, i.e., q is decreasing. If $q_t > q_e$, x is negative, Δq_t is positive, and q is increasing. In either case, q moves away from q_e; the equilibrium under negative overdominance is unstable. A population which is at q_e will remain there; but if $q_t \neq q_e$ initially, it will not approach equilibrium, but go to fixation of *m*, if $q_t > q_e$, or of *M*, if $q_t < q_e$. In an actual population of finite size which is at q_e, sampling variation will tend to move q away from q_e. Once this has happened, fixation of one allele or the other will follow. Hence, negative overdominance equilibria in actual populations would tend to be ephemeral, even if the conditions for them are fulfilled.

Lecture 14 Exercises

1. Assume an African population in a malarial environment, where all sickle cell homozygotes die without reproducing. Normal homozygotes produce only 1/3 as many offspring as heterozygotes, due to the enervating effects of malarial infections. What are the equilibrium allelic proportions, p_e and q_e, in this population? Confirm by showing that $d\bar{W}_t/dq_t = 0$ for these allele frequencies.

2. A World Health Organization team moves into the area and wipes out the malaria-carrying *Anopheles* mosquitoes. Assuming that the population was at equilibrium when this happened, what would you expect p and q to be after one generation of reproduction free from malaria?

3. With negative overdominance, calculate q_e for $h' = 0, 0.1, 0.2, \ldots, 1.0$.

Lecture 15 Mutation

A gene produces its biological effects by serving as a template for the formation of a gene product, usually a polypeptide, which is biologically active (e.g., as an enzyme or a structural protein.) The activity of the gene product is controlled by the sequence of amino acids in its structure; and this, in turn, is controlled by the sequence of the purine and pyrimidine bases in the structure of the gene. We will define a mutation as a permanent, heritable change in the structure of a gene, i.e., a change in the sequence of its bases, whether by rearrangement or deletion of existing bases, or by substitution or addition of new bases. Such a change may effect a change in the structure of the gene product, altering its activity. This in turn may cause a change in the metabolism, morphology and/or behavior of individuals carrying the mutant in homozygous condition (and sometimes in heterozygous condition as well.)

However, some changes in gene structure result in a minor modification of the gene product which does not alter function appreciably. Isozymes, for example, are forms of an enzyme which differ slightly in structure, due to differences in base sequence in the genes which specify their structure (isoalleles). The change in enzyme structure is not sufficient to alter its activity, and only by sophisticated biochemical techniques can the existence of alternate forms of the enzyme be detected (e.g., Truesdale-Mahoney et al. 1981). Some changes in base sequence may not even change the gene product at all. Such changes would, by our definition, still qualify as mutations. The following discussion of the population genetics aspects of mutation will apply to such mutations as well as to gene structure changes with more overt phenotypic effects.

We are considering at present the effects of mutation alone. Mutant genes frequently have deleterious effects on fitness, but we will for the moment ignore any selection consequent upon mutation. Combined effects of mutation and selection will be considered later.

Suppose, then, that we have a gene *M* which mutates repeatedly to an allelic form, *m*. If mutation from *M* to *m* were a unique event which had happened, and could happen, only once, the fate of *m* would be determined by selection and/or chance. But if mutation occurs repeatedly, at a specifiable rate, **u**, per *M* gene per generation, the fate of *m* will depend upon the following considerations.

If **u** is constant, and the frequency of *m* alleles in generation t is q_t,

$$q_{t+1} = q_t + \mathbf{u}p_t = q_t + \mathbf{u} - \mathbf{u}q_t = \mathbf{u} + (1 - \mathbf{u})q_t. \tag{15.1}$$

This is a recursive relationship, true for any two consecutive generations. Consequently,

$$q_{t+2} = \mathbf{u} + (1-\mathbf{u})q_{t+1} = \mathbf{u} + (1-\mathbf{u})[\mathbf{u} + (1-\mathbf{u})q_t]$$
$$= \mathbf{u} + \mathbf{u}(1-\mathbf{u}) + (1-\mathbf{u})^2 q_t;$$
$$q_{t+3} = \mathbf{u} + (1-\mathbf{u})[\mathbf{u} + \mathbf{u}(1-\mathbf{u}) + (1-\mathbf{u})^2 q_t]$$
$$= \mathbf{u} + \mathbf{u}(1-\mathbf{u}) + \mathbf{u}(1-\mathbf{u})^2 + (1-\mathbf{u})^3 q_t;$$
$$q_{n+i} = \mathbf{u} + \mathbf{u}(1-\mathbf{u}) + \mathbf{u}(1-\mathbf{u})^2 + \ldots + \mathbf{u}(1-\mathbf{u})^{n-1} + (1-\mathbf{u})^n q_t. \quad (15.2)$$

In Eqs. (15.2), we can take the sums of the terms in **u**, excluding the final term, $(1-\mathbf{u})^n q_t$, in each case. These sums are

$$\mathbf{u} = \mathbf{u}(1-\mathbf{u})^0 = 1-(1-\mathbf{u});$$
$$\mathbf{u} + \mathbf{u}(1-\mathbf{u}) = 1-[1-\mathbf{u}-\mathbf{u}(1-\mathbf{u})]$$
$$= 1-(1-2\mathbf{u}+\mathbf{u}^2) = 1-(1-\mathbf{u})^2;$$
$$\mathbf{u} + \mathbf{u}(1-\mathbf{u}) + \mathbf{u}(1-\mathbf{u})^2 = 1-[1-\mathbf{u}-\mathbf{u}(1-\mathbf{u})-\mathbf{u}(1-\mathbf{u})^2]$$
$$= 1-(1-3\mathbf{u}+3\mathbf{u}^2-\mathbf{u}^3) = 1-(1-\mathbf{u})^3;$$
$$\mathbf{u} + \mathbf{u}(1-\mathbf{u}) + \ldots + \mathbf{u}(1-\mathbf{u})^{n-1} = 1-(1-\mathbf{u})^n.$$

It therefore follows that

$$q_{t+n} = 1-(1-\mathbf{u})^n + (1-\mathbf{u})^n q_t = 1-(1-q_t)(1-\mathbf{u})^n;$$
$$1-q_{t+n} = (1-q_t)(1-\mathbf{u})^n;$$
$$p_{t+n} = p_t(1-\mathbf{u})^n. \quad (15.3)$$

Equation (15.3) can be used to predict the frequency of *M* genes after n generations of mutation, given that we know p_t and **u**. This prediction takes into account only the effects of mutation. If we find that p_{t+n} differs significantly from the value predicted by Eq. 15.3, the difference may be attributable to other factors, such as selection, migration, or random drift, which may also have been changing allele frequencies during the intervening generations.

We can rearrange Eq. (15.3) into the form

$$(1-\mathbf{u})^n = p_{t+n}/p_t.$$

Expressing this in terms of natural or Napierian logarithms (ln),

$$n \ln(1-\mathbf{u}) = \ln p_t - \ln p_{t+n};$$

if **u** is small,

$$\mathbf{u} \cong (\ln p_t - \ln p_{t+n})/n. \quad (15.4)$$

In theory, if we had two estimates of p, n generations apart in the same population, we could estimate the mutation rate from Eq. (15.4). Again, the assumption is implicit that the only factor effecting changes in allele frequencies is mutation. Because the other factors mentioned above are seldom absent, and because of sampling errors in estimating p, Eq. (15.4) does not really represent a reliable means for estimating **u**.

Fitch and Atchley (1985a) observed 97 loci in ten inbred strains of mice to test methods of reconstructing ancestral relationships among taxa. Their results suggested that either ancestral heterozygosity was greater among these strains than expected, or the mutation rate had been abnormally high. Both Johnson et al. (1985) and Green et al. (1985) object strongly to the suggestion of an elevated mutation rate, though Fitch and Atchley, both in their original paper and in their rebuttal to these comments (Fitch and Atchley 1985b), point out evidence that the mutation rate is not higher than normal. The incident serves as an example of the unreliability of differences in estimates of p as a basis for estimating mutation rate.

Reliable estimates of natural mutation rates for major genes with qualitative effects seldom exceed 10^{-4}. East (1935) suggested that mutation rates for minor genes involved in quantitative inheritance in plants may generally be higher. Mukai (1964), in *Drosophila*, and Bell (1982), in *Tribolium*, have reported evidence of similar phenomena. If a mutation in a gene controlling, say, the structure of an enzyme, causes inactivation of the enzyme, there is a major gene difference between the wild-type and mutant alleles. A minor gene difference between alleles might cause only a slight change in enzyme activity. It is conceivable that any one of a number of changes in base sequence might cause a minor gene difference, whereas a major gene difference could only result from a change in sequence at a few critical sites. In that case, we might expect that minor mutations would occur more frequently than major.

The change per generation in allele frequency due to mutation is

$$\Delta q_t = q_{t+1} - q_t = q_t + \mathbf{u}p_t - q_t = \mathbf{u}p_t. \tag{15.5}$$

The change from generation $t+n$ to generation $t+n+1$ can be written as

$$\begin{aligned}\Delta q_{t+n} &= 1-(1-\mathbf{u})^{n+1}p_t - 1 + (1-\mathbf{u})^n p_t = (1-\mathbf{u})^n p_t(1-1+\mathbf{u}) \\ &= (1-\mathbf{u})^n \mathbf{u}p_t.\end{aligned}$$

Because **u** is very small, $1-\mathbf{u}$ is nearly 1; then $(1-\mathbf{u})^n$ is close to 1 unless n is very large. For example, if $\mathbf{u}=10^{-6}$, $1-\mathbf{u}=0.999999$. This fraction must be raised to the 6-th power before it rounds to 0.9999 rather than to 1.0000. Thus, for $n<6$, setting $(1-\mathbf{u})^n$ equal to 1 introduces an error of less than 0.01%. For small values of n,

$$\Delta q_{t+n} \cong \mathbf{u}p_t = \Delta q_t; \tag{15.6}$$

the change in allele frequency per generation is virtually constant, at least over periods of a few generations.

However, $(1-\mathbf{u})$ is, in fact, less than 1, though only slightly less in most cases. Therefore, $(1-\mathbf{u})^n$ will go to zero as n becomes very large. The ultimate consequence of mutation from *M* to *m* will be the elimination of *M* genes from the population. All *M* genes will eventually mutate to *m*; there is no intermediate equilibrium point. From Eq. (15.5), as long as **u** is greater than 0, Δq can only be 0 when p is 0. The approach to p=0 will be asymptotic, because the number of *M* genes lost per generation will decrease as p becomes smaller and smaller.

It is, of course, also possible to have reverse mutation from *m* to *M*, as well as forward mutation from *M* to *m*. Suppose that, in addition to forward mutation at rate **u**, we also have reverse mutation at rate **v**. Then

$$\Delta q_t = \mathbf{u}p_t - \mathbf{v}q_t. \tag{15.7}$$

Since forward mutation increases q, and back mutation decreases it, an equilibrium is possible when the gain and loss of *m* alleles exactly balance. Allele frequencies at equilibrium can be calculated:

$$\mathbf{u}p_e = \mathbf{u} - \mathbf{u}q_e = \mathbf{v}q_e;$$

$$\mathbf{u} = (\mathbf{u}+\mathbf{v})q_e;$$

$$q_e = \mathbf{u}/(\mathbf{u}+\mathbf{v});\ p_e = \mathbf{v}/(\mathbf{u}+\mathbf{v}). \tag{15.8}$$

This is a stable equilibrium. If $q_t \neq q_e$, because $\mathbf{u} = (\mathbf{u}+\mathbf{v})q_e$,

$$\Delta q_t = \mathbf{u} - \mathbf{u}q_t - \mathbf{v}q_t = (\mathbf{u}+\mathbf{v})q_e - (\mathbf{u}+\mathbf{v})q_t$$
$$= (\mathbf{u}+\mathbf{v})(q_e - q_t).$$

Then Δq_t will be positive if $q_t < q_e$, negative if $q_t > q_e$. In either case, q moves toward q_e.

The equilibrium frequency of *m* genes will be 0.5 if forward and reverse mutation rates are equal. Generally, however, forward are likely to be much higher than reverse mutation rates. Any of a number of changes in the structure of the gene *M* might cause changes in product structure producing qualitatively, perhaps even quantitatively, similar changes in the function of that product. Hence, recurrent mutation to *m* might not represent a single specific change in gene structure occurring repeatedly, but different structure changes, to any one of a collection of mutant forms having similar effects on gene product function.

Reverse mutation, the repair of a particular mutant form to the functionally normal *M* gene, would require a very specific change in the structure of that mutant form. The specific reverse change would be much less likely than an essentially random forward mutation. A difference of two orders of magnitude between the forward and reverse rates is not unusual.

With forward mutation rates greater than reverse, we might expect to find many populations with mutants outnumbering the "wild-type" gene. Suppose, for example, that $\mathbf{u} = 10^{-5}$ and $\mathbf{v} = 10^{-7}$. Then at equilibrium

$$q_e = 10^{-5}/(10^{-5}+10^{-7}) = (100)(10^{-7})/(101)(10^{-7}) = 0.99;$$

$$p_e = 0.01.$$

99% of the genes in the population should be *m*. In fact, we seldom see such situations in nature. Equilibrium due to this balance, although theoretically possible, does not occur very often.

The "wild-type" gene produces a product to which the organism as a whole is adapted. The product of the mutant fails to fulfill a metabolic need, and the mutant is often at a selective disadvantage. Selection eliminates *m* genes while mutation renews the supply, and we might well expect to find some balance between mutation and selection determining allele frequencies. We must next discuss equilibria due to a balance between mutation and selection.

Lecture 15 Exercises

1. Given that $p_t = 0.8$, estimate

   a) p_{t+3} and p_{t+5}, if $\mathbf{u} = 5 \times 10^{-6}$;

   b) $\mathbf{u}$, if $p_{t+10} = 0.799862$;

   c) the number of generations needed to reach $p = 0.7995$, with $\mathbf{u} = 5 \times 10^{-6}$.

2. Calculate q_e when $\mathbf{u} = 5 \times 10^{-5}$, $\mathbf{v} = 5 \times 10^{-6}$.

Lecture 16 Interaction of Mutation with Selection

In any real population, mutation will be taking place. There will also be natural selection, and, in the case of domestic plants or animals, there may be artificial selection as well. Selection, either artificial or natural, tends to increase the frequency of one allele at a given locus until that allele is in the preponderance. Mutation tends to decrease the frequency of the preponderant allele. Thus, it is almost inevitable that selection and mutation will have opposite effects on allele frequencies.

Suppose, for example, that we have *M* mutating to *m*, the mutation rate being **u**, along with selection against *mm* homozygotes and, to a lesser degree, *Mm* heterozygotes. Selection tends to decrease the frequency of *m*; mutation, to increase it. Furthermore, as this allele becomes rarer, the change per generation in allele frequency due to selection decreases; at the same time, the change due to mutation increases, because the frequency of *M* genes is increasing. Under these circumstances, one might well expect to find some equilibrium point at which the decrease due to selection is exactly balanced by the increase due to mutation. The present lecture will attempt to quantify this concept.

Suppose, then, that we have a population segregating for alleles *M* and *m*, whose respective frequencies in generation t are p_t and q_t. There is selection against the *m* allele, so that relative fitnesses are:

$$w_{11}: w_{12}: w_{22} = 1: 1-hs: 1-s.$$

There is also recurrent mutation from *M* to *m*, at rate **u**.

The change in q between generations t and t+1 due to selection will be, from Eq. (13.6),

$$\Delta_S q_t = -sp_t q_t (h - 2hq_t + q_t)/(1 - 2hsp_t q_t - sq_t^2).$$

This change is negative for all values of $q_t > 0$, and would lead eventually to elimination of *m* and fixation of *M*.

The change in q due to mutation between generations t and t+1 will be

$$\Delta_u q_t = \mathbf{u} p_t,$$

from Eq. (15.5). This change, positive for all values of $p_t > 0$, would lead eventually to the fixation of *m* and the elimination of *M*.

The combined effect of mutation and selection is approximately the sum of these two change functions:

$$\Delta_C q_t \cong \Delta_u q_t + \Delta_S q_t \qquad (16.1)$$

The sum of the change functions only approximates the combined change, because mutation alters q_t in $\Delta_S q_t$ and selection alters it in $\Delta_u q_t$. If both effects are small, their sum is a reasonably good approximation of the combined change.

The population will be at equilibrium under the combined effects of mutation and selection when $\Delta_C q_e = 0$; that is, when the positive effects of mutation and the negative effects of selection exactly balance. When $\Delta_C q_e = 0$, $\Delta_u q_t = -\Delta_S q_t$; solving this equation for q_e should give us the allele frequency at equilibrium. If $\Delta_C q_e = 0$,

$$\mathbf{u}p_e = sp_e q_e(h - 2hq_e + q_e)/(1 - 2hsp_e q_e - sq_e^2);$$

$$\mathbf{u} = sq_e(h - 2hq_e + q_e)/(1 - 2hsp_e q_e - sq_e^2);$$

$$q_e = \mathbf{u}(1 - 2hsp_e q_e - sq_e^2)/s(h - 2hq_e + q_e). \tag{16.2}$$

The solution again depends on the dominance relations of *M* and *m*, with regard to their effect on fitness, which determines the value of h. If the fitness effects of *m* are completely recessive, so that heterozygotes are as fit as the *MM* homozygotes, $h = 0$, and

$$q_e = \mathbf{u}(1 - sq_e^2)/sq_e;$$

$$q_e^2 = \mathbf{u}(1 - sq_e^2)/s. \tag{16.3}$$

We can approximate the solution by taking $1 - sq_e^2 \cong 1$. Then

$$q_e = (\mathbf{u}/s)^{1/2}. \tag{16.4a}$$

Because $\mathbf{u}$ is likely to be much smaller than s, Eq. (16.4a) indicates that q_e will be very small. [Curiously, Eq. (16.4a) is exact, although based on the approximate Eq. (16.1). When $h = 0$, the error introduced by adding mutation and selection effects is cancelled by the $1 - sq_e^2$ approximation. See Crow 1983.]

This relationship accounts for the persistence, in human populations, of some severely deleterious recessive genetic diseases, such as phenylketonuria, cystic fibrosis and albinism. Phenylketonuria (PKU), for instance, is a very severe disease with a high neonatal death rate; phenylketonurics who survive infancy usually show severe mental retardation and elevated mortality rates at all subsequent ages. They are very unlikely to reproduce (Thompson and Thompson 1980). Heterozygotes, on the other hand, show neither disease symptoms nor any significant reduction in fitness. The disease is thus an example of a situation in which $s = 1$ and $h = 0$.

Despite the selection coefficient of 1 against homozygotes, the disease persists, with an incidence that is nearly constant at about 7 cases per 100,000 live births (7×10^{-5}). We can assume this incidence to be the frequency of PKU homozygotes at equilibrium; thus,

$$q_e^2 = 7 \times 10^{-5};$$

$$q_e = 8.4 \times 10^{-3}.$$

Applying Eq. (16.3), if

$$q_e = 8.4 \times 10^{-3} = (\mathbf{u}/s)^{1/2},$$

with s = 1,

$$\mathbf{u} = 7 \times 10^{-5}.$$

This appears to be a reasonable mutation rate for the gene causing PKU. We can assume that PKU is indeed kept in the population as a result of mutation replacing genes that are eliminated by selection.

If a deleterious mutation is completely dominant to the normal allele with regard to fitness, so that heterozygotes are under as strong a selection disadvantage as homozygotes, h = 1, and Eq. (16.2) gives us

$$q_e = \mathbf{u}(1 - 2sq_e + sq_e^2)/s(1 - q_e).$$

We found that q_e was quite small when h = 0 and there was no selection against the heterozygote. Here, where the heterozygote as well as the homozygote are selected against, q_e should be even smaller. The expressions in parentheses in the above equation should therefore both be approximately equal to 1, and

$$q_e \cong \mathbf{u}/s \qquad (16.4b)$$

is a reasonable approximation.

It is difficult to find a good example of a dominant deleterious mutation which is held in the human population by an equilibrium between mutation and selection. This is partly because, if s = 1 in Eq. (16.4b), the equilibrium frequency of the mutant gene will be the mutation rate. Mutant genes disappear as rapidly as they are formed; the only mutant genes in the population at a given time will be those that have mutated in the current generation. Any mutations from earlier generations will have been eliminated by selection. Even though individuals heterozygous for the mutant gene will show the mutant phenotype, this phenotype will be very rare.

One of the most important indications of genetic causation of a disease or abnormality, especially in human data, is the fact that genetic diseases often show a familial tendency, a tendency for the disease to recur within the same family. In PKU, for example, the frequency of the disease in the general population is 7×10^{-5}. But if a normal couple have already produced one PKU child, both parents must be heterozygotes, and the probability that a second child of that couple will also have the disease is 0.25, while a normal sib of the afflicted child will be more likely than not to carry the gene. Collateral relatives of the couple will also be more likely to carry PKU than the general public. Thus, PKU exhibits a familial tendency, which originally helped to identify the disease as genetically caused.

With a dominant deleterious mutation, if s = 1, the appearance of one mutant child is very likely to have been caused by a mutation occurring in that child's own genotype, and may not signal an increased risk of disease for later sibs. Thus, the effects of the mutation might well not be recognized as being of genetic origin at all.

If the fitness effects of the *m* gene are partially dominant, h lies between 0 and 1. There is selection against the heterozygote, but it is less severe than that against the mutant homozygote. One might then expect q_e to lie between the values for the dominant and recessive mutations:

$$\mathbf{u}/s < q_e < (\mathbf{u}/s)^{1/2},$$

as is indeed the case. An approximate solution of Eq. (16.2), if we assume that hq_e and sq_e are both virtually zero, is

$$q_e \cong \mathbf{u}/hs. \tag{16.4c}$$

Clearly, this is greater than $\mathbf{u}/s$, because

$$\mathbf{u}/hs = (1/h)(\mathbf{u}/s);$$

h being a fraction less than 1, $1/h > 1$. At the same time,

$$(\mathbf{u}/s)^{1/2} = (\mathbf{u}/s)/(\mathbf{u}/s)^{1/2} = (s/\mathbf{u})^{1/2}(\mathbf{u}/s),$$

Since s is usually several orders of magnitude greater than $\mathbf{u}$, $(s/\mathbf{u})^{1/2}$ will be larger than $1/h$ unless h approaches 0. But Eq. (16.4c) is not valid if h is very small, that is, if the deleterious gene is almost completely recessive.

The effect of reverse mutation has been ignored in the above formulations. Reverse mutation will, of course, act to reduce q_e, but its effect will generally be negligible. The effect of reverse mutation on the *m* allele frequency at equilibrium would be $-\mathbf{v}q_e$. As we have seen, both q_e and $\mathbf{v}$ are small, and their product will be miniscule. The correction to q_e for reverse mutation effects would be smaller than the sampling error of any estimates of q_e we might make.

The effect of mutation in a situation with positive overdominance would be to increase slightly the frequency of *m*, and decrease that of *M*, alleles. Reverse mutation would, of course, have the opposite effect. Then

$$q_e' = q_e + \mathbf{u}q_e - \mathbf{v}q_e.$$

The difference between this solution and the one given in Eq. (14.3) is negligible.

Lecture 16 Exercises

1. Calculate q_e for the following values of s and h. Let $\mathbf{u} = 10^{-6}$, and use the appropriate formulae developed in the text.

$s = 1, 1/4$

$h = 0, 1/2, 1$

2. Evaluate the case of PKU, using the more exact formula given by Eq. (16.3), and compare to the result in the text, which uses Eq.(16.4).

3. For the case where $s = 1/4$, $h = 1/2$, calculate the more exact value of q_e in terms of $\mathbf{u}$, then evaluate q_e for $\mathbf{u} = 10^{-6}$.

Lecture 17 Migration

The word migration in common speech refers to the movement of people or animals from one geographical location to another. Seasonal migration among birds is a well-known phenomenon. Many species migrate, with the coming of spring, to breeding grounds in northern latitudes, retreating in autumn to southern climes. (As I write these words, a honking flock of wild geese is passing overhead, migrating northward.)

Another type of migration involves the permanent movement of whole populations to new territories. Human inhabitants of the North American continent should be particularly cognizant of this type of migration, since we are all descendants of migrants from other continents. Even the American Indian descends from migrants coming out of Asia in prehistoric times. However, neither this sort of Völkerwanderung, nor seasonal migration, will be the subject of this lecture.

Our attention will instead be concentrated on the movement of individuals between populations, which is yet a third type of migration. An emigrant leaving a population takes with him his genotype; an immigrant imports his with himself. If emigrants and immigrants differ in genotype, there will be a net change in genotypic and perhaps in allelic frequencies in the population as a result of migration.

Let us suppose, then, that we have a population, S, segregating for alleles *A* and *a* at frequencies p_t and q_t, respectively, in generation t before migration takes place. In each generation, some fraction, m, of the members of the population emigrate, being replaced by an equal number of immigrants from outside population S. Migration takes place before the population reproduces, so that immigrants reproduce within the population, emigrants do not. The emigrants are drawn at random from the population; in generation t, the *a* allele frequency among them is therefore q_t. We will designate the frequency of *a* alleles among the immigrants in generation t as q^*_t.

The frequency of *a* alleles in generation t+1 zygotes will therefore be

$$q_{t+1} = q_t - mq_t + mq^*_t = q_t - m(q_t - q^*_t)$$

so that

$$\Delta q_t = -m(q_t - q^*_t). \tag{17.1}$$

Obviously, if either m or $q_t - q^*_t$ is zero, Δq_t will be zero, and there will be no change in allele frequencies due to migration. If m is 0, there is no migration; this is a trivial equilibrium solution. The equilibrium situation we are interested in is that which arises when $q_t - q^*_t = 0$. This says simply that if the frequency of *a* alleles in emigrants and in immigrants is the same, migration will not change allele frequency in the population. The *a* alleles removed from S by emigrants will be replaced by *a* alleles brought in by immigrants.

The population will approach equilibrium over time if q^*_t remains constant. Consider

$$q_{t+1}-q^*_t=q_t-m(q_t-q^*_t)-q^*_t=q_t-q^*_t-m(q_t-q^*_t)$$
$$=(1-m)(q_t-q^*_t) < (q_t-q^*_t). \qquad (17.2)$$

The difference between q in emigrants and immigrants is less in generation t+1 than it was in generation t, and it will continue to decrease as time goes on.

$$q_{t+2}-q^*_t=(1-m)(q_{t+1}-q^*_t)=(1-m)^2(q_t-q^*_t);$$
$$q_{t+n}-q^*_t=(1-m)^n(q_t-q^*_t). \qquad (17.3)$$

Because $1-m$ is a fraction less than 1, $(1-m)^n$ goes to 0 as n becomes large. The eventual result of migration will be to make the *a* frequency within the population equal to q^*_t, at which point equilibrium is reached.

What is the source of the immigrants "outside the population S"? Because *omne vivum ex ovo*, these immigrants cannot arise in the void by spontaneous generation. They must come from another population or populations existing in geographical proximity to S. S can be thought of as a member of a cluster of populations; emigrants out of other populations in the cluster supply the immigrants into S. If these migrants are chosen at random from the other populations, q^*_t is the average *a* frequency among them.

Emigrants from S may simply disappear into the void, but it is more reasonable to assume that they enter other populations in the cluster, presumably more or less at random. In doing so, they will affect allele frequencies in those other populations. As a result, q^*_t is not in fact constant. It is decreased by the frequency of *a* in immigrants into, and increased by that in emigrants out of, the study population. If k is the total number of populations in the cluster, including S, and all populations are similar in size, the change in q^*_t from generation t to generation t+1 is

$$\Delta q^*_t=-m(q^*_t-q_t)/(k-1). \qquad (17.4)$$

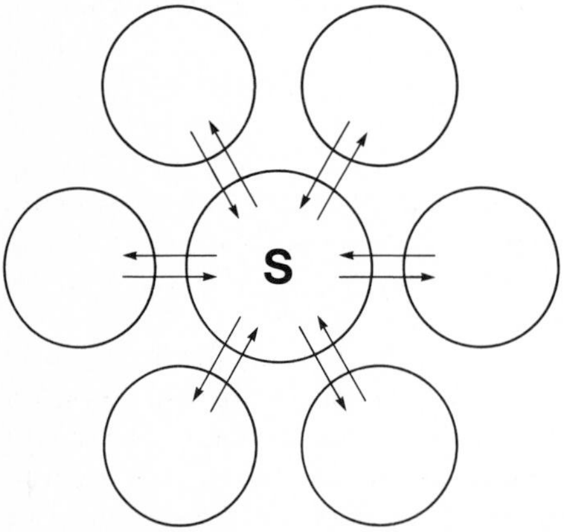

Fig. 17.1. A diagrammatic representation of migration. Population S exchanges migrants with the surrounding populations in the cluster. Immigrants enter population S from the other populations; emigrants from S enter the other populations. The *a* frequency in emigrants is q_t, in immigrants q^*_t, in generation t

The change in q^* per generation is slight. Emigrants out of S are absorbed among a total number of individuals $k-1$ times as great as the number in population S, so that their effect is only $1/(k-1)$ as great as that of the equal number of immigrants into S. Our original formulation in terms of a constant q^* is not seriously in error, unless m is large or k small.

What change does occur in q^* is opposite in sign to the change in q. As we saw from Eq. (17.3), q tends to approach q^*; from Eq. (17.4), q^* tends to approach q. The approach to equilibrium ($q^* = q$) is accelerated by the change in q^*.

We could have designated any one of the populations in the cluster as S. In each, the same process is occurring, immigrants entering from all the other populations, and emigrants leaving to join the others. Thus, each population's *a* allele frequency is moving toward the average of all the rest. The eventual result is that all populations stabilize at a common overall average allele frequency.

Note that m must be a reasonably small fraction. If there are only two populations, and $m = 1/2$, we have the population mixture described in Lecture 6, which is not likely to be described as migration. If m is large in a multi-population cluster situation, the populations within the cluster lose identity. Instead of a cluster of populations, we have a single superpopulation.

Let us look again at the change in frequency of *a* due to migration in the study population. From Eq. (17.1),

$$\Delta_G q_t = -m(q_t - q^*_t) = -mq_t + mq^*_t.$$

We have labeled this Δ with a subscript G, to emphasize that it is due to migration. We can multiply this equation by $p^*_t + q^*_t$ $(= 1)$ without changing its value:

$$\begin{aligned}\Delta_G q_t &= (-mq_t + mq^*_t)(p^*_t + q^*_t) \\ &= mp^*_t q^*_t + mq^{*2}_t - mp^*_t q_t - mq^*_t q_t \\ &= mq^*_t(p^*_t + q^*_t - q_t) - mp^*_t q_t \\ &= mq^*_t p_t - mp^*_t q_t. \qquad (17.5)\end{aligned}$$

Equation (17.5) expresses the change in q due to migration in a form similar to that used in Eq. (15.7) to express the change due to forward and reverse mutation. We can therefore write the joint effect of migration and of mutation in both directions as

$$\Delta_J q_t = \Delta_G q_t + \Delta_u q_t = (mq^*_t + \mathbf{u})p_t - (mp^*_t + \mathbf{v})q_t. \qquad (17.6)$$

Either mutation or migration can replace a *A* allele with a *a*, or vice versa. As far as allele frequencies are concerned, it is immaterial which process brings about the replacement. Of course, migration rates can be much higher than mutation rates, so that migration is potentially a more powerful source of change in alelle frequency than mutation.

Migration, nonetheless, is generally an unimportant factor for the plant or animal breeder, simply because it usually can be controlled so easily. In only a few situations, such as open-pollinated crops in the field, or range cattle, is the introduction of foreign genetic material of any importance. The breeder tends to work with closed populations, or crosses among such populations. Ideally, he would prefer to control every mating which takes place among his stocks, and he can usually come fairly close to realizing that ideal.

It should also be noted that q_e under migration is likely to be much nearer to 1/2 than the balance between forward and reverse mutation. Suppose, for example, that we have two populations of equal size, one with a very high and the other a very low frequency of *a*. Mutual migration between the two will tend to bring both to an equilibrium q_e close to 1/2.

We have assumed above that migration took place at random, in the sense that emigrants from a population were a random sample of the population members. We can, however, easily imagine situations in which this might not be the case. Suppose that we have contiguous populations occupying environments that differ slightly. If one genotype were attracted or repelled by an environmental factor that differed between the two environments, we might expect a differential migration of that genotype from one population to the other.

For example, suppose that we have two populations, X and Y, of an insect species which lives by sucking plant sap. These insects normally infest a wide range of plant species found prolifically throughout the contiguous territories of the two populations. There exists, however, in both populations a mutant gene, *e*, homozygotes for which have a defect in the proboscis with which the sap is obtained. As a result of this defect, *ee* homozygotes can only obtain sap from a limited number of plant species, and these are more common in the territory infested by population Y, because of slight differences in climate and soil conditions.

Mutant *ee* homozygotes hatched in population X territory must fly far afield to find suitable food plants. In the process, some may cross into population Y territory. Once having done so, they are unlikely to return to population X territory, because their source of food is readily available where they are. For the same reason, *ee* homozygotes hatched in population Y territory are unlikely to cross into the territory of population X. *EE* and *Ee* genotypes hatched in either territory are unlikely to cross boundaries at all; a small proportion will do so, but migrants in one direction are likely to be balanced by migrants in the opposite direction.

If the frequencies of *E* and *e* are p_{Xt}, q_{Xt} in generation t of population X, and p_{Yt}, q_{Yt} in contemporaneous generation t of Y, we can assume genotypic ratios in the two populations before migration to be

$$p_{X,t}^2 : 2p_{X,t}q_{X,t} : q_{X,t}^2;$$

$$p_{Y,t}^2 : 2p_{Y,t}q_{Y,t} : q_{Y,t}^2.$$

Migration takes place before reproduction, so that at reproduction, the frequency of *ee* individuals in population X has been decreased by the proportion, m, of *ee* emigrants. The distribution of genotypes in population X at reproduction has therefore become

$$p_{X,t}^2 : 2p_{X,t}q_{X,t} : (1-m)q_{X,t}^2.$$

Clearly, the effect of differential migration on population X has been equivalent to a reduction in the fitness of *ee* homozygotes. The migration rate is equivalent to a selection coefficient, and the relative "fitnesses" of the genotypes in population X can be written as

$$w_{11}: w_{12}: w_{22} = 1: 1: 1-m.$$

The general selection equation can then be applied; the eventual result of differential migration will be the elimination of *e* alleles from population X.

In population Y, the effect of *ee* immigrants from X will be to increase the frequency of *e*. The genotypic distribution at reproduction will be

$$p_{Y,t}^2: 2p_{Y,t}q_{Y,t}: q_{Y,t}^2 + mq_{X,t}^2.$$

The frequency of *ee* homozygotes in population Y at reproduction can be written as

$$q_{Y,t}^2(1+mq_{X,t}^2/q_{Y,t}^2) = (1+mb_t)q_{Y,t}^2,$$

where b_t is the ratio of *ee* frequency in population X to that in population Y in generation t. This ratio will obviously go to zero as the frequency of *ee* in X goes to zero. If there are no *ee* in population X, they cannot influence the frequency of *ee* in population Y by migrating into the latter population.

The relative "fitnesses" in population Y are

$$w_{11}: w_{12}: w_{22} = 1: 1: 1+mb_t.$$

In this form we can apply the general selection equation. The selection coefficient, in this case, enters as a positive rather than a negative value. Alternatively, we can convert to a form in which the fitness of *ee* is 1, and those of the other two genotypes are less than 1.

Thus, nonrandom migration can be handled like selection. Differential emigration from one population reduces the contribution of one genotype to the gametes that will form the next generation as effectively as if the individuals of that genotype died. Differential immigration into the other population increases the contribution of one genotype, thereby decreasing the relative contribution of the other genotypes.

The example we have presented is oversimplified. Nonetheless, it may serve to show how nonrandom migration can act to change populations.

Lecture 17 Exercises

1. We have a group of five contiguous populations, A, B, C, D and E. These populations interchange individuals every generation, with a migration rate of 0.1. At generation t, *a* allele frequencies are 0.1, 0.3, 0.5, 0.7 and 0.9 in populations A through E, respectively. Calculate q_{t+1} for each population. What will the equilibrium q value be for population A?

2. Three contiguous populations interchange members on the following basis: 1/3 of the individuals born in A move to B and 1/3 to C each generation; similarly, 1/3 of the individuals born in B move to A and 1/3 to C, and 1/3 born in C move to B and 1/3 to A. If allele frequencies in the three populations are q_{At}, q_{Bt} and q_{Ct} in generation t, what will these frequencies be in generation t+1?

Part IV Dispersive Forces

Lecture 18 The Finite Population Model

In previous lectures, we have assumed infinitely large populations, and infinitely large samples from these populations. This permitted us to equate frequencies with their underlying probabilities. For example, the probability of drawing a *A* gamete in a sample from generation t of a population was p_t, where p_t was the frequency of *A* in the infinite population. Because the sample was of infinite size, we could assume that p_t was also the frequency of *A* gametes drawn.

We must now consider populations and samples of finite size, in which frequencies are no longer necessarily equal to the corresponding probabilities. Any finite population can be considered as a sample drawn from some theoretical infinite population. In the process of drawing the sample, sampling error is introduced. In a finite sample drawn from the infinite population of gametes, the probability of a *A* gamete is still p_t, but the frequency of *A* gametes drawn will be p_t plus or minus sampling error.

The amount of sampling error depends on the size of the sample. If the population, although finite, is very large, sampling error may be negligibly small. But if the sample is very small, sampling error may be so great that the parameter is totally unrecognizable from the statistic. In such a sample, for example, no *A* gamete may be drawn, even though p_t is a substantial fraction.

Sampling error can affect the genetic constitution of a population at three levels. The genotypes which take part in reproduction are a sample from among the genotypes available to reproduce. The gametes which unite to form the next generation are a sample from the gametes produced by the reproducing genotypes, and the genotypes actually formed by the gametes which unite are a sample from the possible genotypic combinations. Thus, sampling errors can affect the distributions of reproducing genotypes, gametes, and offspring genotypes. In a large population, each of these distributions will differ only slightly from its expected value; but in a small population, each is subject to considerable error. Errors at one level of sampling may tend to cancel those at another, but it is also possible that the errors accumulate across levels.

To make predictions about the effects of finite population size, we must quantify the effects of sampling error. To try to do so, we will employ the following model (from Falconer 1960, 1981). Figure 18.1 may help to illustrate this model.

We start with a base population of infinite size, which is reproducing under Hardy-Weinberg conditions: that is, the organism is diploid, segregating for alleles *A* and *a* at a single autosomal locus, and mating at random with no mutation, migration, or selection. The frequencies of *A* and *a* in this base population are p_0 and q_0, respectively.

We introduce finite population size by dividing the base population into an infinite number of lines, each of N individuals. The N individuals in a given line

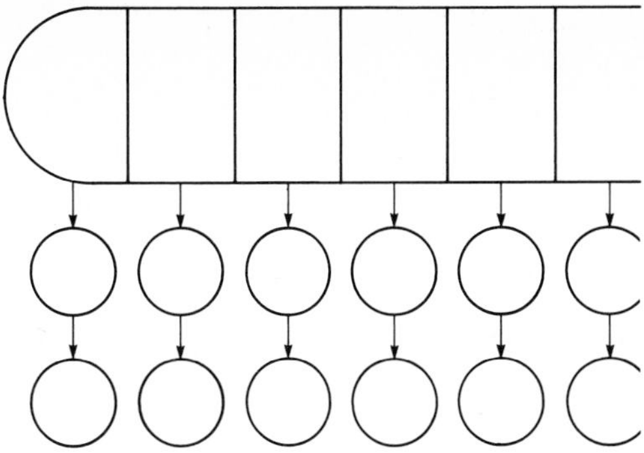

Fig. 18.1. The finite population model. The original infinite population is broken up into an infinite number of samples of size N, each drawn at random, and each forming the base generation for a line of size N

are chosen at random without replacement from the base population. Each individual in the base population will appear in one, and only one, of the lines.

We then allow the members of each line to mate together at random. No matings can occur between members of one line and those of any other. The matings in any one line could in theory produce an infinite population of possible offspring, from which we obtain a sample of N individuals at random. This procedure accomplishes sampling of gametes and of zygotic combinations of gametes. The i-th sample chosen from the infinite population will be numbered generation 1 of the i-th line; their offspring, generation 2, etc. Generation 2 individuals are mated randomly within line to produce generation 3, subject to the same sampling rules as applied to the previous generation.

Each line is a finite population, and the fate of these lines will illustrate the effects of finite population size on the genetic constitution of successive generations. These effects can be described in terms of three parameters: inbreeding, or the frequency of identical gene pairs; genic drift, or the variation of allelic frequencies among lines; and genotypic drift, or the variation in genotypic frequencies among lines. We will describe the effects in terms of each of these parameters; but it should be emphasized that the three are not different phenomena. All three parameters describe the same phenomenon, although in different terms. One or another may be the most convenient description for a particular purpose, but all three parameters are changing simultaneously, and, as we shall see, the parameters are interconvertible. Given the inbreeding coefficient, we can estimate genic and genotypic variances among lines, and vice versa.

Lecture 19 Inbreeding

To consider the effects of finite population size in terms of inbreeding, we define inbreeding as the union of a pair of gametes which carry genes at a given locus which are identical by descent. The inbreeding coefficient, F, measures the probability of such a union.

The concept of identity by descent (Malecot 1948) rests upon the fact that the genes carried by any individual are copies of the genes carried by his parents. One of the two genes at any given locus in a diploid is a copy of a gene carried by his sire, the other of one of his dam's genes. He in turn will transmit to each of his offspring a copy of either his paternal or his maternal gene at that locus, and so ad infinitum. All genes in generation t are thus copies of generation t-1 genes, and those are copies of genes from generation t-2, etc.

Any two copies of the same gene are said to be identical by descent (or merely identical; genic identity can arise only by descent). If an individual passes on to two of his offspring copies of, say, his paternal gene, the two copies are identical to one another, to the parent's paternal gene, and to the gene in the parent's sire from which all are descended. They are also all identical to any other copies of the same ancestral gene that may exist in the population.

Identity can carry over many generations. If B is a lineal descendant of A, B may carry an identical copy of one of A's genes, no matter how many generations may intervene between A and B. If C also descends from A, C can also carry a copy of the same gene from A, and thus B and C can carry identical genes. B, as A's descendant, is a lineal relative of A. B and C, both descendants of A, are collateral relatives. Two individuals can carry identical genes in common only if they are either lineal or collateral relatives.

If a pair of relatives, lineal or collateral, mate, there is some probability that their offspring, D, will receive identical genes from its two parents. This probability is the inbreeding coefficient of D, F_D. D is said to be inbred if $F_D > 0$, which can only occur if D's parents carry an identical gene in common. Thus, our definition of inbreeding equates to mating between two individuals that have some probability of carrying identical genes in common.

Webster's dictionary defines inbreeding as the mating of related individuals, and this is probably a more familiar definition than the one we have given in terms of genic identity. The two definitions converge, however. If two individuals are related, they have some chance of carrying identical genes in common; to carry identical genes, they must be related.

The inbreeding coefficient of an individual can be calculated from its pedigree. If its parents are unrelated, F for that individual must be 0. If its parents are related, the individual's F is the probability that it receives identical genes from both parents. That probability diminishes as the distance in generations between his parents and their common ancestor increases, because there is a probability in each intervening generation that a descendant of the common ancestor will pass on a copy of its other gene. We will discuss the calculation of individual inbreeding coefficients in detail later.

It should be emphasized that F is a probability. Any given individual does or does not carry identical genes at a given locus. If we had an infinite population of D's, each an offspring of B x C, F_D would be the frequency of D's in the population that do carry identical genes at a particular locus.

Although F is a per locus probability the same probability, of course, applies to each locus in a given individual, because the pedigree is the same for every locus. Because there is a very large number of loci in any individual, F_D approximately measures the proportion of loci in D which carry a pair of identical genes. Linkage reduces the accuracy of this approximation; if one locus carries identical genes, other loci linked closely with it are also likely to carry identical genes. Loci linked to a locus carrying nonidentical genes are also likely to carry genes that are not identical.

In the present lecture, we will not be too concerned with individual inbreeding coefficients. We could not evaluate all possible coefficients, even for a single line with small N. If we could do so, we could not predict which individuals, and which F values, would be realized in the random sampling process. We can, however, and we will, estimate F_{it}, the average inbreeding coefficient of individuals in generation t of line i.

Since gametes unite at random in each generation of line i, F_{it} is simply the probability that a pair of random gametes from generation t-1 carry identical genes. Suppose that we have, at an autosomal locus in generation t-l of line i, 2N genes $A_1, A_2, \ldots, A_{2N}$ (where the subscripts refer to identities, rather than alleles). If we choose two gametes at random from t-1, and both carry copies of the same gene, say A_j, the zygote resulting from this union will be inbred. The probability of choosing a gamete carrying A_j is 1/2N on either of the two independent draws needed to provide the pair of gametes. The probability of drawing A_j twice is therefore $(1/2N)^2$. Because the same probability obtains for each of the 2N genes available from generation t-1, the probability of drawing two gametes, both carrying copies of the same generation t-1 gene, is $2N(1/2N)^2 = 1/2N$.

This is true whether these two gametes unite together, both appearing in the same offspring individual, or whether each unites with a separate gamete, and appears in a different offspring individual. Thus the probability of identity is the same between two genes in the same individual or in different individuals.

The probability of drawing gametes carrying genes A_j and A_{j*} from generation t-1 is therefore $1-(1/2N)$. But these two genes, although different genes in generation t-1, can be identical copies of the same gene in some prior generation. The probability that two genes in generation t-1 are identical is F_{it-1}. Therefore, the total probability of uniting a pair of gametes carrying identical genes is

$$F_{it} = (1/2N) + (1 - 1/2N)F_{it-1}. \qquad (19.1)$$

The first term in equation 19.1 is the probability of drawing, from generation t-1, multiple copies of the same gene in that generation. We can think of this as "new" inbreeding, arising *de novo* in generation t. The second term is the prob-

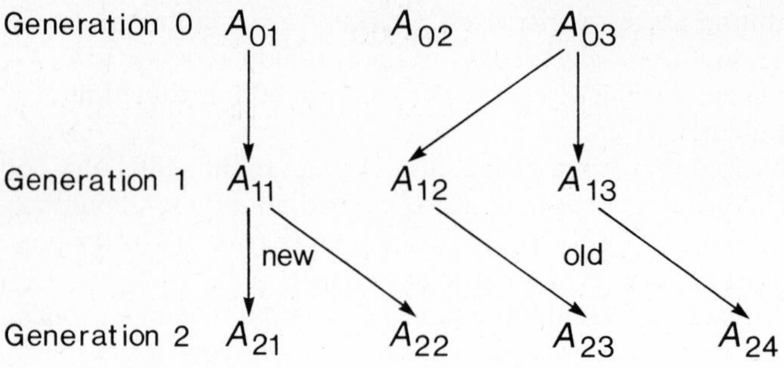

Fig. 19.1. Inbreeding. Genes A_{01}, A_{02}, and A_{03} in generation 0 are all nonidentical. Gene A_{11} in generation 1 is a copy of gene A_{01} in the previous generation; genes A_{12} and A_{13} are both copies of gene A_{03}. In generation 2, genes A_{21} and A_{22} are both copies of A_{11} from generation 1. Gene A_{23} copies A_{12}, and A_{24} copies A_{13}; they are therefore both copies of A_{02}. The identity between A_{21} and A_{22} represents new inbreeding in generation 2; that between A_{23} and A_{24}, old inbreeding

ability that two different genes that are identical in t-1 are both copied in generation t. This represents the retention in generation t of "old" inbreeding that arose in earlier generations. Figure 19.1 may be helpful in understanding the distinction between new and old inbreeding. If N is small, 1/2N will be relatively large, and a good deal of new inbreeding will arise in generation t. At the same time, however, relatively little of the old inbreeding will be retained. If N is large, there will be little new inbreeding, but a great portion of the old inbreeding will be retained. If N fluctuates from generation to generation, inbreeding arising in generations when N is small will be retained in those when N is large. Expanding a small population to a very large size, while it will retard further new inbreeding, will tend to preserve the inbreeding acquired when the population was small.

If we let $X=1-(1/2N)$, so that $1/2N=1-X$, we can write Eq. (19.1) in the form

$$\begin{aligned} F_{it} &= 1-X+XF_{it-1} = 1-X+X(1-X+XF_{it-2}) \\ &= 1-X+X-X^2+X^2F_{it-2} = 1-X^2(1-F_{it-2}) \\ &= 1-X^t(1-F_{i0}) = 1-[1-(1/2N)]^t(1-F_{i0}) \end{aligned} \tag{19.2}$$

Thus, we see that inbreeding in generation t of line i is a function of t, N, and the initial inbreeding in the base population. But in our model, all lines stem from the same base population, so that F_{i0} is simply F_0, the inbreeding in the base population, common to all lines. Then F_{it} is the same for all lines, and we can write it as simply F_t.

This is not to say that the frequency of identical genes is the same for all lines. The probability is the same, but since each line represents a sampling process, the frequencies of identical genes vary from line to line.

The base population is infinitely large, and the probability that any two genes in it are both copies of the same gene in the preceding generation is infinitesimally small. New inbreeding cannot occur. All inbreeding existing in the previous generation is, of course, retained, but if the previous generation was also infinitely large, new inbreeding could not have arisen then either. If we assume that the population has always been infinitely large, no inbreeding can ever have arisen, and $F_0 = 0$. Every one of the infinite number of genes in the base population is a separate identity; no two are identical.

Assuming $F_0 = 0$ may seem rather presumptuous. It has, however, a certain amount of practical reasonableness. We could assume any value for F_0 between 0 and 1. If we assumed $F_0 = 1$, all base generation genes would be identical; so would all genes in each generation of each line, and the whole calculation would be meaningless. But it is not very likely that all base generation genes are identical.

There seems to be no good reason to set F_0 equal to any particular value between 0 and 1. If we set it equal to 0, however, we at least have the consolation that, even if this is inaccurate, F_t will reflect accurately any inbreeding accumulated since generation 0.

In addition, setting $F_0 = 0$ has a parallel in what is done when inbreeding coefficients are calculated for individuals from pedigree information. It is seldom possible to trace a pedigree back for more than a few generations. We will eventually, and often rather quickly, come to a generation of ancestors whose relationships are unknown and about whose inbreeding coefficients we must make some assumption. It is common practice to assume that these individuals are unrelated and not inbred. This is equivalent to assuming $F_0 = 0$ in our base population. We will therefore make this assumption in our investigation of the effects of finite population size.

In our earlier investigations of the effects of systematic forces and of multiple alleles, sex linkage, etc., we ignored inbreeding. We were justified in doing so, because we assumed, implicitly or explicitly, that we were dealing with an infinitely large population. In a later lecture, we will consider the interactions between inbreeding and some of these other effects.

Lecture 19 Exercises

1. Calculate F_1 through F_5 for lines of sizes $N = 2$, 10, and 100, from Eq. (19.1). Note and record the amounts of new inbreeding arising and old inbreeding retained in each generation. Assume $F_0 = 0$.

2. Recalculate the inbreeding coefficients for a line with $N = 2$ in generations 1 and 2, suddenly increased to 100 in generation 3 and subsequent generations.

3. Using Eq. (19.2), calculate the inbreeding coefficient in generations 5, 10, and 20 for the lines of Exercise 1.

Lecture 20 The Development of Inbreeding

Let us examine the development of inbreeding in a finite line of our model population. From Eq. (19.1), we can calculate the change in F from generation t to generation t+1 of line i as

$$\begin{aligned} F_{it+1}-F_{it} &= (1/2N)+[1-(1/2N)]F_{it}-F_{it} \\ &= (1/2N)+F_{it}-(1/2N)F_{it}-F_{it} \\ &= (1/2N)(1-F_{it}). \end{aligned} \tag{20.1}$$

F_{it} is a probability, with a value that must always lie between 0 and 1, so that $1-F_{it}$ must be 0 or positive. The term 1/2N likewise can never be negative. Therefore, Eq. (20.1) can never be negative; it must always be positive unless it is 0. F must increase or remain constant.

The change in F will be 0 if the line is infinitely large, because then $1/2N=0$. We have already argued that F must be constant in an infinitely large line, since new inbreeding can never arise. For finite N, $1/2N > 0$; therefore, the change in F can only be 0 if $F_{it}=1$. In a finite line, F_{it} constantly increases until it reaches 1. Because F_{it} is the probability that a pair of genes drawn at random from generation t of line i will be identical, any pair of genes drawn from one line will be identical if $F_{it}=1$. In other words, when $F_{it}=1$, all genes in line i are identical; the line is completely inbred.

We can also conclude from Eq. (19.2) that F_{it} will continually increase to its maximum value of 1. Assuming $F_0=0$, Eq. (19.2) can be written as

$$F_{it}=1-[1-(1/2N)]^t. \tag{20.2}$$

For finite N, $1-(1/2N)$ is a fraction between 1/2 and 1. As the power, t, to which this fraction is raised increases, the fraction becomes smaller, and F_{it} correspondingly becomes larger. As t becomes very large, the fraction goes to 0, and F goes to 1.

Panmixia is a fancy word for random mating. We define the panmictic index of line i in generation t as the measure of the "noninbreeding" remaining in the line at that generation, $P_{it}=1-F_{it}$. Thus, P_{it} measures the probability that a random pair of genes drawn from generation t of line i are not identical. The panmictic index can be written in the recursive form:

$$\begin{aligned} P_{it} &= 1-F_{it}=1-(1/2N)-[1-(1/2N)]F_{it-1} \\ &= [1-(1/2N)](1-F_{it-1})=[1-(1/2N)]P_{it-1} \\ &= [1-(1/2N)][1-(1/2N)]P_{it-2}=[1-(1/2N)]^t P_{i0}. \end{aligned} \tag{20.3}$$

Again, $[1-(1/2N)]^t$ decreases as t increases, eventually going to 0, so that P_{it} goes to 0 as t becomes very large. We can restate the conclusion drawn above, that inbreeding continually increases to 1, by saying that the panmictic index continually decreases to 0.

By analogy with our definition of Δq_t in earlier lectures, we should define ΔF_{it} as Eq. (20.1), the change in F between generations t and t+1 of line i. We carefully refrained from referring to this equation as ΔF_{it}, because by convention, ΔF_{it} is defined as the new inbreeding in generation t:

$$\Delta F_{it} = 1/2N. \tag{20.4}$$

If N is constant, ΔF_{it} is a constant for all generations of line i, and in our model population for all lines.

Substituting ΔF_{it} thus defined into Eq. (20.1),

$$F_{it+1} - F_{it} = \Delta F_{it}(1 - F_{it}).$$

After rearranging,

$$\Delta F_{it} = (F_{it+1} - F_{it})/(1 - F_{it}); \tag{20.5a}$$

which we can also write in terms of the panmictic index:

$$\Delta F_{it} = (1 - P_{it+1} - 1 + P_{it})/P_{it} = (P_{it+1} - P_{it})/P_{it}. \tag{20.5b}$$

ΔF_{it} measures the increase in inbreeding (or the decrease in noninbreeding) from generation t to generation t+1 of line i as a ratio to the noninbreeding remaining in generation t of line i. This ratio is constant, which means that the amount of inbreeding gained per generation decreases as F increases and P decreases. If, for example, $\Delta F_{it} = 0.25$ (N=2), with $F_0 = 0$ ($P_0 = 1$), P_{i1} will be 0.75; a quarter of the panmixis is lost in the first generation. But in the next generation, the loss will only be 0.25(0.75)=0.1875; in the following generation, 0.1406; etc. Thus, F approaches unity asymptotically.

The concept of genic identity specifies that two genes are identical if both are copies of the same gene in a common ancestor. If we assume no mutation, both must be the same allele. Since inbreeding increases the frequency of identical genes, it increases homozygosity. In a completely inbred line, with F=1, all genes are identical, and therefore all are the same allele.

Every gene in generation t of line i is identical with, and therefore the same allele as, some gene in the original base population. When two gametes carrying identical genes unite, the resulting individual must be homozygous, as the two identical genes must be the same allele. The probability that the ancestral gene was *A* is p_0, the frequency of *A* alleles in the base population. Therefore, the probab ility that uniting identical genes in generation t gives rise to a *AA* homozygote is p_0; to a *aa* homozygote, q_0. The probability that such a union gives rise to *Aa* is, of course, 0.

Two genes in generation t of line i that are identical to different base population genes are not identical to one another. The probabilities that two random base population genes are both *A*, one *A* and one *a*, or both *a* are, of course, p_0^2: $2p_0q_0$: q_0^2. Therefore, the probability that uniting two genes that are not identical in generation t of line i will give rise to *AA* is p_0^2, etc.

The frequency of a union of gametes carrying genes identical in generation t of line i is F_{it}, of gametes carrying genes not identical, $1-F_{it}$. It follows that the probabilities of *AA*, *Aa*, and *aa* genotypes in generation t+1 will be

$$D_{it+1}=p_0F_{it}+p_0^2(1-F_{it})=p_0^2+(p_0-p_0^2)F_{it}$$
$$=p_0^2+p_0q_0F_{it},$$
$$2H_{it+1}=0F_{it}+2p_0q_0(1-F_{it})=2p_0q_0-2p_0q_0F_{it}, \text{ and}$$
$$R_{it+1}=q_0F_{it}+q_0^2(1-F_{it})=q_0^2+p_0q_0F_{it}, \quad (20.6)$$

respectively. Equations (20.6) show that the probabilities of homozygotes increase as F increases. The probability of a heterozygote must correspondingly decrease, so that the sum of the three probabilities remains equal to 1; the terms in $p_0q_0F_{it}$ cancel.

Eventually F will reach 1. When this happens, the probabilities of the three genotypes will be, from Eqs. (20.6),

$$p_0^2+p_0q_0=p_0(p_0+q_0)=p_0,$$
$$2p_0q_0-2p_0q_0=0,$$
$$q_0^2+p_0q_0=q_0.$$

In a completely inbred population, there are no heterozygotes; *AA* and *aa* homozygotes appear with probabilities equal to the frequencies of the two alleles in the base population.

Our model population, as a whole, contains an infinite number of lines. Equations (20.6) represent the frequencies of the various genotypes across all lines. They also represent the expected or average frequencies in any one line. Each line will, of course, deviate from this average or expectation, due to sampling variation. At $F=1$, no line can exist in which there are p_0 *AA*, q_0 *aa*, and 0 *Aa*. Each line mates within itself randomly, and random mating within such a line would immediately produce heterozygotes. The situation is analogous to saying that the average size of completed families in a human population is, say, 2.3 children; no family with exactly 2.3 children can actually occur. This statement must be interpreted to mean that some families have 2 children or less, some 3 or more, so that on the average, the number of children is 2.3.

Similarly, the average distribution

$$AA\colon Aa\colon aa=p_0\colon 0\colon q_0$$

must be interpreted to mean that some lines have one type of homozygote, others have the other. With complete inbreeding, the frequency of lines homozygous for *A* is p_0, for *a*, q_0. No line will carry any *Aa*.

Thus, inbreeding results in increased homozygosity of lines. Equations (20.6) can be used to determine the distribution of genotypes across all lines in generation t. This will be the expected frequency for any one line, but each line will vary from the expected value.

The breeder will, of course, never have an infinite number of lines of finite size, all derived from a common base population, at his disposal. He will have a few such lines at best; more frequently, he may have only one. His best a priori guess as to the genotypic constitution of that line is given by Eqs. (20.6). He knows, however, that the actual genotypic distribution in his line in generation t will vary from the result given by these equations. He knows for certain that, at complete inbreeding, his line will not contain p_0 *AA* and q_0 *aa* individuals, but will be entirely homozygous for one allele or the other. In his completely inbred line, p will be either 0 or 1. Along the way, then, p in his line must have departed from p_0, through random sampling of gametes and genotypes. We must, therefore, attempt to describe how allelic and genotypic frequencies change due to sampling.

Lecture 20 Exercises

1. An infinite population is broken up into lines of N individuals each. What are ΔF and F_{10} if N=2? N= 10?

2. If the original frequencies of *A* and *a* were 0.6 and 0.4, respectively, what are the expected frequencies of the three genotypes in generation 10, for N=2 and for N=10?

Lecture 21 Genic Drift

The frequencies of alleles and of genotypes will change from generation to generation in a finite line, due to sampling error. These changes are referred to in the literature as "random genetic drift." We will consider changes in allelic and in genotypic frequencies separately, referring to the former as genic, and the latter as genotypic, drift.

We initiate line i in our model population by drawing a random sample of N individuals from the base population. Since each individual carries a random pair of genes, this represents a random sample of 2N genes. Only rarely will the frequency of *a* genes in the i-th sample, q_{i1}, be exactly equal to the frequency, q_0, in the base population. In most lines, q_{i1} will be greater or less than q_0. However, the average of the line frequencies, $\bar{q}_{i1}$, will be equal to q_0. Every gene in the base population will appear in one and only one of the lines. Therefore, the frequency of *a* across all lines must be the same as in the base population. Positive and negative deviations of the q_{i1} must cancel one another over the infinite population of lines.

The possible offspring of the N individuals in generation 1 of line i, mated at random, form an infinite population, in which the frequency of *a* alleles is q_{i1}. We sample N individuals, 2N genes, at random from this population. Again, the frequency of *a* alleles in the sample will only rarely be exactly q_{i1}; most q_{i2} will be greater or less than the corresponding q_{i1}. However, the sum of the deviations of the q_{i2}'s from their respective q_{i1} values, over an infinite number of lines, is zero; positive and negative deviations cancel. Therefore the average frequency of *a* across all lines in generation 2 is the same as the average in the preceding generation; $\bar{q}_{i2} = \bar{q}_{i1} = q_0$. Because this is true in each successive generation,

$$\bar{q}_{it} = \ldots = \bar{q}_{i2} = \bar{q}_{i1} = q_0; \qquad (21.1)$$

the average allele frequency across all lines remains constant, equal to the frequency in the base generation.

Nevertheless, q_{it} for a particular line i will change, and will change independently of changes in any other line. Therefore, the lines will come to differ in allele frequencies. We can predict neither q_{it} for any particular line i, nor the difference in allele frequencies, $(q_{it} - q_{i*t})$, for any particular pair of lines i and i*. We can, however, predict the variance among the q_{it} across all lines, and can use this variance to describe changes in allele frequencies.

Because the genes in generation 1 of line i are a sample from the genes of the base generation, the variance among the q_{i1} is the binomial variance:

$$V_1(q_i) = p_0 q_0 / 2N.$$

In subsequent generations, the variance among the q_i becomes compounded. In each line in each generation, we sample from the genes in the previous generation. The process corresponds to a series of Lexian trials (Crow 1954), from which the variance after t generations is

$$\begin{aligned} V_t(q_i) &= p_0q_0/2N + (1-1/2N)V_{t-1}(q_i) \\ &= p_0q_0/2N + XV_{t-1}(q_i), \end{aligned} \tag{21.2}$$

if we again let $X = 1-1/2N$. This relationship is recursive, so that

$$\begin{aligned} V_t(q_i) &= p_0q_0/2N + Xp_0q_0/2N + X^2V_{t-2}(q_i) \\ &= p_0q_0/2N + Xp_0q_0/2N + X^2p_0q_0/2N + \ldots + X^{t-1}p_0q_0/2N \\ &\quad + X^tV_{t-t}(q_i)\,. \end{aligned}$$

But $V_{t-t}(q_i)$ must be the variance prior to any sampling, that is, in the undivided base population, which is patently 0. Therefore,

$$V_t(q_i) = (p_0q_0/2N)(1 + X + X^2 + \ldots + X^{t-1}).$$

The sum of the series $1 + X + X^2 + \ldots + X^{t-1}$ is $(1-X^t)/(1-X)$. Substituting this into the previous equation,

$$V_t(q_i) = p_0q_0(1-X^t)/2N(1-X).$$

But

$$2N(1-X) = 2N(1-1+1/2N) = 2N(1/2N) = 1,$$

and therefore,

$$V_t(q_i) = p_0q_0(1-X^t). \tag{21.3}$$

From Eq. (21.3), $V_t(q_i)$ is a function of allele frequencies in the base population and of N, and can be evaluated directly from these values. We do not have to develop the variance in generation t by calculating the variance in each successive generation, starting from generation 1.

The change in $V_t(q_i)$ per generation can be evaluated from Eq. (21.3):

$$\begin{aligned} V_t(q_i) - V_{t-1}(q_i) &= (1-X^t)p_0q_0 - (1-X^{t-1})p_0q_0 \\ &= p_0q_0(1-X^t-1+X^{t-1}) = (X^{t-1}-X^t)p_0q_0. \end{aligned} \tag{21.4}$$

But $X = 1-1/2N$ is a fraction less than 1, so that $X^{t-1} > X^t$.

Therefore, $X^{t-1} - X^t > 0$. The change function, Eq. (21.4), is positive; $V_t(q_i)$ increases as t increases.

Thus, the effect of genic drift is to increase the variance among the q_i values, even though $\bar{q}_i$ remains constant; that is to say, the *a* allele frequencies in the various lines become more disparate with time.

As t increases, $X^t = (1-1/2N)^t$ decreases, going to 0 in the limit as t becomes infinitely large. If 0 is the limiting value of X^t, the limiting value of $1-X^t$ is 1. Therefore, $V_t(q_i)$ has a limiting value also:

$$V_L(q_i) = p_0q_0,$$

the value which the variance assumes when $1-X^t = 1$.

The variance of the q_i has a limit because q_i itself has limits. Being a frequency, q_i can never be greater than 1, nor less than 0. The maximum difference in allele frequencies between lines i and i_* occurs when $q_i = 1$, and $q_{i*} = 0$.

In any line which still contains both *A* and *a* alleles in generation t, the frequency of *a* can either increase or decrease due to sampling in generation t+1. If, however, $q_{it} = 1$, further sampling can no longer change allele frequency in line i. All genes in the line are *a*, and any sample must have q_{it+1} equal to 1. Similarly, if q_{i*t} is 0, all genes in line i_* are *A*, and q_{i*t+1} must be 0 in any sample. The values 0 and 1 are absorbing barriers for q_i.

This being the case, the eventual fate of any line will be to become fixed for either *A* or *a*. Each line, changing in allele frequency in each successive generation, will eventually by chance reach either $q_i = 1$ or $q_i = 0$, and will thereafter contain only one allele.

The average q_{it}, however, will remain equal to q_0, even when all lines are fixed. This can only be true if p_0 of the lines become fixed for *A*, q_0 for *a*. This will therefore be the limiting distribution of lines.

If in the limit there are p_0 lines with $q_{iL} = 0$, and q_0 lines with $q_{iL} = 1$,

$$\bar{q}_{iL} = p_0(0) + q_0(1) = q_0.$$

The variance of the q_{iL} is

$$\begin{aligned} V_L(q_i) &= p_0(0-q_0)^2 + q_0(1-q_0)^2 \\ &= p_0q_0^2 + q_0p_0^2 = p_0q_0. \end{aligned}$$

This result agrees with the limiting value of $V_t(q_i)$ we previously derived from Eq. (21.3).

Consideration of genic drift therefore leads us to the conclusion that, in the limit, there will be p_0 lines fixed for *A* and q_0 fixed for *a*. Wahlund (1928) showed that the expected frequencies of genotypes in generation t of line i will be

$$\begin{aligned} D_{it} &= p_0^2 + V_t(q_i); \\ 2H_{it} &= 2p_0q_0 - 2V_t(q_i); \\ R_{it} &= q_0^2 + V_t(q_i). \end{aligned} \tag{21.5}$$

But the limiting value of $V_t(q_i)$ is

$$V_L(q_i) = p_0q_0,$$

so that equations 21.5 will, in the limit, go to

$$\begin{aligned} D_{iL} &= p_0^2 + p_0q_0 = p_0; \\ 2H_{iL} &= 2p_0q_0 - 2p_0q_0 = 0; \\ R_{iL} &= q_0^2 + p_0q_0 = q_0. \end{aligned}$$

The parallels between genic drift and inbreeding are obvious. Both predict that the eventual fate of the population of finite lines will be the fixation of *A* in p_0, and of *a* in q_0, of the lines. The analogy between Eqs. (21.2) and (19.1) has already been noted. In each case, new inbreeding (variation) induced by sampling in generation t is added to a portion of the preexisting variation (inbreeding) retained from previous generations. In fact, either equation can be written in terms of either F_t or $V_t(q_i)$. From Eq. (20.2), we can write

$$F_t = 1 - X^t = 1 - (1 - 1/2N)^t;$$

and from Eq. (21.3),

$$V_t(q_i) = p_0 q_0 (1 - X^t).$$

It follows that

$$V_t(q_i) = p_0 q_0 F_t. \qquad (21.6)$$

Substituting from Eq. (21.6) into Wahlund's formulae [Eqs.(21.5)] gives us

$$D_{i,L} = p_0^2 + p_0 q_0 F_t,$$

and so forth. Thus, Wahlund's formulae are equivalent to the formulae given in Eq. (20.6). Finally, the limiting value of $V_t(q_0)$ is $p_0 q_0$. Then, to be consistent with Eq. (21.6), the limiting value of F_t must be

$$F_L = V_L(q_0)/p_0 q_0 = p_0 q_0/p_0 q_0 = 1,$$

which, as we already know, is correct.

Thus, inbreeding and genic drift are not separate phenomena, but merely different ways of describing the same phenomenon, the effect of finite line size.

Lecture 21 Exercises

1. In the original infinite population, let $p_0 = 0.6$, $q_0 = 0.4$. The population is broken up into lines of size $N = 5$. Calculate $\bar{q}_{it}$ and $V_t(q_i)$ for $t = 1, 2, ..., 5$. Check your calculations by directly calculating $V_5(q_i)$ from Eq. (21.3).

2. What is the limiting value of $V_t(q_i)$?

Lecture 22 Genotypic Drift

In the previous lecture, we described the consequences of drawing 2N random genes from each generation in each line. The sampling procedure actually applied, of course, was to sample N diploid individuals, N genotypes, in each generation of each line, rather than 2N genes per se.

If we have a random sample of K observations on variable X, we can calculate the arithmetic mean of the sample,

$$\bar{X} = \Sigma X_j/K.$$

We define the expected value of X, E(X), as the limiting value which $\bar{X}$ will approach as K becomes infinitely large. The expected value of X is the mean of the $\bar{X}$ values from an infinite number of samples.

In a sample of N random genotypes from generation t of line i, the observed frequency of *aa* is R_{it}. If we drew K samples from generation t of line i, each sample containing N random genotypes, the frequency in the j-th sample would be $(R_{it})_j$. The set of $(R_{it})_j$ values would itself be a sample of K observations on variable R_{it}; the mean of this sample would be

$$\bar{R}_{it} = \Sigma(R_{it})_j/K.$$

and $E(R_{it})$ would be the limiting value of $\bar{R}_{it}$.

If p_{it}, q_{it}, are the allelic frequencies among gametes drawn at random from generation t of line i, the possible offspring obtained by uniting these gametes at random form an infinite population with genotypic ratio

$$p_{it}^2 : 2p_{it}q_{it} : q_{it}^2.$$

Therefore, q_{it}^2 is the probability that an individual from the population will be *aa*. Our sample of N random genotypes in t is drawn from this population. If N were infinite, this probability would also be the frequency of *aa* in the sample. Since the population is finite, R_{it} will not be equal to q_{it}^2, due to sampling error. However, q_{it}^2 will be the expected value of R_{it}, that is, the mean of the R_{it} values from a infinite series of samples. Therefore,

$$E(R_{it}) = q_{it}^2. \qquad (22.1)$$

In fact, we do not have an infinite number of samples from any one line in generation t. We have an infinite number of samples, one from each line, and the relationship in Eq. (22.1) holds true for every line. In some lines, R_{it} will be greater than q_{it}^2 for that line, in others it will be less. Positive and negative deviations will cancel, so that across all lines,

$$E(R_{it}) = E(q_{it}^2). \qquad (22.2)$$

$E(q_{it}^2)$ is the mean q_{it}^2 across all lines, the limiting value of $\Sigma q_{it}^2/K$. This is the sum of the squared *a* allele frequencies, divided by K. The square of the sum of the allele frequencies, divided by K, is quite a different quantity:

$$(\bar{q}_{it})^2 = [\Sigma(q_{it}/K)]^2;$$

taken across all lines,

$$[E(q_{it})]^2 = q_0^2. \tag{22.3}$$

From the product-moment definition of a variance,

$$V_t(q_i) = \Sigma(q_{it} - \bar{q}_{it})^2/K = [\Sigma q_{it}^2 - (\Sigma q_{it})^2/K]/K$$
$$= \Sigma q_{it}^2/K - (\Sigma q_{it})^2/K^2. \tag{22.4}$$

But from Eq. (22.2), $E(q_{it}^2) = E(R_{it})$ is the limiting value of Σq_{it}^2; and from Eq. (22.3), the limiting value of $(\Sigma q_{it})^2/K^2$ is $[E(q_{it})]^2 = q_0^2$. Because we are taking $V_t(q_i)$ across an infinite number of lines, we can substitute expected values into Eq. (22.4):

$$V_t(q_i) = E(R_{it}) - q_0^2.$$

We can rearrange this equation to define $E(R_{it+1})$ in terms of $V_t(q_i)$ and q_0. By similar reasoning, we can define $E(D_{it+1})$ and $E(2H_{it+1})$ also in terms of these two variables:

$$E(D_{it}) = p_0^2 + V_t(q_i);$$
$$E(2H_{it}) = 2p_0q_0 - 2V_t(q_i);$$
$$E(R_{it}) = q_0^2 + V_t(q_i). \tag{22.5}$$

These equations should by now have a perhaps dreary familiarity; they are, of course, Wahlund's formulae, Eqs. (21.5). As t becomes very large, $V_t(q_i)$ goes to p_0q_0; the genotypic ratio therefore goes to

$$AA\colon Aa\colon aa = p_0\colon 0\colon q_0.$$

Genotypic drift arises from the same source, leads to the same result, and is interconvertibly comeasurable with inbreeding and genic drift; it is simply a third way of describing the same phenomenon.

All lines eventually go to fixation; but the fixation process is gradual, and it is reasonable to suppose that at various stages in the process, some lines have become fixed while others continue to segregate. Thus, Eqs. (22.5) provide the expected frequencies of the three genotypes after t generations of breeding with finite line size, but no single line may have these frequencies; they are an average over all lines. Wright (1952) derived formulae for the distribution of fixed and segregating lines in generation t:

fixed for *A*: $p_0 - 3p_0q_0(1 - 1/2N)^t$;

fixed for *a*: $q_0 - 3p_0q_0(1 - 1/2N)^t$;

segregating: $6p_0q_0(1 - 1/2N)^t$. (22.7)

As t becomes very large, these equations go to p_0, q_0, and 0, respectively. [Equations (22.7) are only approximate; they represent the first terms in series expressions. Later terms in the series are, however, negligibly small. See Crow and Kimura 1970.]

Kimura (1955) has pointed out that the fixation process is not adequately described by Wright's equations. The q_{i1} values, the initial frequencies of *a* in the various lines, tend to be clustered fairly closely around q_0, unless N is very small. Over the next several generations, the q_{it} tend to spread across the range 0 to 1, eventually assuming a more or less rectangular distribution with all values of q from 0 to 1 becoming equally probable. Until this rectangular distribution has been achieved, fewer lines will become fixed than predicted by Wright's equations. The equations do represent a reasonable estimate of fixation frequencies after the rectangular distribution has been reached. The time required to reach rectangularity is a function of q_0 and N.

Lecture 22 Exercises

1. The population in the Lecture 21 exercise has $p_0 = 0.6$, $q_0 = 0.4$, $N = 5$. Calculate the expected frequencies of the three genotypes after 5 generations, and after 20 generations, in this population.

2. For the same population, what proportion of lines will be fixed after 20 generations?

Lecture 23 Effective Breeding Size

In describing the effects of finite population size, we used as a model a Hardy-Weinberg population of infinite size, divided at random into an infinite number of lines of finite size N. We then related our measures of the effect of finite size to N; e.g., the rate of inbreeding was $\Delta F = 1/2N$.

The plant or animal breeder, or the biologist describing a natural population, will never have the situation envisaged in our model population. He will have at most a few finite lines, and these lines will be affected by systematic forces simultaneously with the effects of finite size. But even aside from these differences between the model population and the actual population which the breeder has available, a variety of differences will be likely to exist which can be encompassed in the term population structure. As a result of these differences, the actual population will not mate randomly in complete conformity with the random mating of the model population. The rate of inbreeding in the actual population will not, as a result, be 1/2N.

Wright (1931, 1938) showed, however, that the expected rate of inbreeding in an actual line could be related to N_e, the effective breeding size of the actual line; $\Delta F = 1/2N_e$. An actual line with N breeding individuals per generation will inbreed at a rate equivalent to the rate in a model line of N_e breeders. The relationship between N_e, the effective breeding size of the line, and N, the actual number of breeders, is determined by the breeding structure of the line. For many of the commonly occurring differences in structure between the model and actual populations, N_e can be calculated from N.

In organisms with which animal breeders usually deal, sexes are separate, and self-fertilization, which can occur in the model population, is impossible. Higher plants frequently produce gametes of both sexes on the same individual; even then, self-sterility or the breeder's intervention may prevent selfing. We would expect the elimination of selfing to decrease the rate of inbreeding, and that is indeed the case. Preventing selfing increases N_e relative to N,

$$N_e \cong N + 1/2, \tag{23.1}$$

so that the inbreeding rate is decreased:

$$\Delta F \cong 1/(2N+1).$$

Obviously, the correction for the elimination of selfing is of little importance if N is even moderately large. If, for example, N is 10, ΔF in the model population will be 0.05; in a population without selfing, ΔF will be 0.0476, a decrease of less than 5%. This is because, under random mating, the probability that two gametes from the same individual will unite is small when N is large, and ignoring this small probability will have little effect.

If N is small, on the other hand, eliminating selfing may have a considerable effect. With $N=2$, ΔF is 0.25 in the presence of selfing, and only 0.20 if selfing is prevented, a 20% decrease in the inbreeding rate.

It should be noted that, with sexes separate, N is the sum of the number of males plus the number of females used in breeding:

$$N=N_m+N_f.$$

We have tacitly assumed, above, that $N_m=N_f$; but this may not always be the case.

In domestic animals, the amount of product obtained from a herd or flock is often dependent on the number of females, whether the product be eggs from a laying flock of fowl, milk from a dairy herd, or offspring to be grown for meat from hogs or beef cattle. The total number of animals that can be raised and maintained is limited by space and labor costs, so that it is obviously more profitable to reduce the proportion of males and increase that of females. Only as many males need be maintained as required to ensure fertility of matings. With artifical insemination from a central stud, of course, this can be carried to the extreme; individual dairy farmers need keep no bull at all. A few bulls kept by the stud impregnate cows by the thousands in herds over a wide area.

In some natural species, monogamy (one male to one female) is the rule. The snow goose, *Chen hyperborea*, forms monogamous matings that last the lifetime of the mated pair, to cite only one example. But in many other species, polygyny, one male mated to several females, is the more normal situation. In many herd species, and even in man's closest relatives, the monkeys and apes, the herd may include a number of males, but only a few "dominant males" will be sexually active; these will sire virtually all offspring of the herd. Polyandry, one female mated to a number of males, is a much less common phenomenon.

As polygamy, usually polygyny, is common in both domestic and natural populations, we must explore its effects on N_e. If we have, say, 1 male and 5 females mating, rather than 3 of each sex, we must expect a greater proportion of half sib offspring. This should increase the frequency of identical genes; the inbreeding rate will increase, N_e will decrease. Because male and female contribute equally to each offspring, we might expect N_e to be a function of the mean of N_m and N_f, and indeed it is. But because N_e appears in the denominator of ΔF, the appropriate mean is not the arithmetic mean, but the harmonic mean of N_m and N_f.

The harmonic mean of a variable X is

$$H^*(X_1, X_2, ..., X_n)=[(1/n)(1/X_1+1/X_2+...+1/X_n)]^{-1};$$

that is, it is the reciprocal of the average reciprocal of the X's. Applying this to N_m and N_f,

$$H^*(N_m, N_f)=[(1/2)(1/N_m+1/N_f)]^{-1}.$$

This harmonic mean represents the effective number in each sex. Because N includes both sexes, this mean must be doubled. Multiplying by 2 is equivalent to multiplying by $(1/2)^{-1}$, so that

$$N_e=2H^*(N_m, N_f)=[(1/4)(1/N_m+1/N_f)]^{-1}. \tag{23.2}$$

The rate of inbreeding is then (Wright, 1938)

$$\Delta F = 1/2N_e = 1/[(1/8)(1/N_m + N_f)]^{-1} = (1/8N_m + 1/8N_f).$$

A peculiarity of the harmonic mean is that smaller values of X are of greater importance in determining $H^*(X_i)$ than larger values. The arithmetic mean of 10 and 100 is

$$\bar{X} = (10 + 100)/2 = 55;$$

but

$$H^*(X_i) = [(1/2)(1/10 + 1/100)]^{-1} = 18.18,$$

much nearer to 10 than to 100. The number of individuals of the less numerous sex is more important in determining the inbreeding rate than the number of the more numerous. Consider, for example, a population of 10 males and 100 females, in which

$$\Delta F = 1/80 + 1/800 = 0.0138.$$

If we halve the number of females,

$$\Delta F = 1/80 + 1/400 = 0.015,$$

increasing the rate of inbreeding by about 9%. But if we instead halve the number of males,

$$\Delta F = 1/40 + 1/800 = 0.0262;$$

the rate of inbreeding increases by about 90%.

Calculation of N_e from Eq. (23.2) ignores the correction [Eq. (23.1)] for avoiding selfing. Unequal numbers in the sexes implies separate sexes, and the absence of selfing, and 1/2 should be added to Eq. (23.2) to provide the correct actual value of N_e. But this correction would have little effect in most cases, and can usually be ignored. In the case of 10 males and 100 females, for example, the inbreeding rate would be 0.0138 if the selfing correction is ignored, and 0.0136 if it is included. The difference is less than 2%.

We assumed a constant N in our model population. This may not be true in practice; natural populations in particular are subject to large changes in population size due to changes in environmental conditions. The harmonic mean can again be applied to provide an average N_e over a series of generations during which N varies (Wright 1938):

$$\bar{N}_e = [(1/t)(1/N_1 + 1/N_2 + \ldots + 1/N_t)]^{-1}. \tag{23.3}$$

Properly, the N_i entered into this equation should be corrected for inequality of sexes and, if small, for suppression of selfing.

Again, the smallest N_i have the greatest effect on N_e and on ΔF. We saw earlier that new inbreeding accumulated during generations when N is small tends to be retained when N becomes larger; Eq. (23.3) indicates the same result.

In our model for the effects of finite population size, we assumed random mating within each line. That is, we assumed that we drew random pairs of individuals from the population of possible genotypes produced by that line, and obtained a single offspring from each pair. The probability of drawing the same pair of mates more than once, so as to produce full sib offspring, was relatively small.

In animal breeding, however, we frequently encounter full sib families. In the pig, and in laboratory rodents, a single mating produces a litter of full sibs. In birds, a whole clutch of eggs from a given dam will be fertilized by sperm from the same sire. Even if a species typically produces only one offspring per birth, repeat matings may be made, and multiple births occasionally take place. Thus, the population of offspring from which the animal breeder must choose mates for the next generation is often a population of full sib families. Wright (1938) presented a formula to correct N_e for the effect of choosing mates from among full sib families, assuming single pair matings, that is, equal numbers of males and females:

$$N_e = (4N-2)/(2+\sigma_k^2), \tag{23.4}$$

where k_i is the number of mates drawn from the i-th family, and σ_k^2 is the variance among the k_i.

If equal numbers of offspring from each full sib family are available for choice as mates, the k_i will approximate a Poisson distribution, in which the variance is equal to the mean ($\sigma_k^2 = \bar{k}$). If line size is to remain constant, on the average each family must supply two mates for the next generation. Therefore

$$\bar{k} = 2 = \sigma_k^2.$$

Thus, from Eq. (23.4), if k has a Poisson distribution,

$$N_e = (4N-2)/(2+2) = N-0.5.$$

If a line consists of N/2 single pair matings; if each mated pair produces the same number of full sib offspring available for choice as mates in the next generation; and if N mates are then chosen at random, the k_i will approximate a Poisson distribution, and N_e will be nearly equal to N.

If families differ in size, however, as is usually the case, there will be a tendency for overrepresentation of the larger families, and underrepresentation of the smaller, among the individuals chosen as mates to produce the next generation. $\bar{k}$ will still be equal to 2, but σ_k^2 will be greater than 2, $2+\sigma_k^2 > 4$, and $N_e < N-0.5$.

At the same time, Eq. (23.4) suggests that a breeder can intervene to increase N_e and decrease ΔF by reducing σ_k^2 to less than 2. If, in fact, the breeder could choose exactly 2 mates from each full sib family, $\sigma_k^2 = 0$, $N_e = 2N-1$, and ΔF would be cut nearly in half.

It is seldom possible for the breeder to attain the maximum reduction in the inbreeding rate. Usually, there will be some infertile matings, and there may be one or two that have only one offspring available for mating. The breeder should then choose two mates per family so far as possible, then make up the remainder of the N mates he needs by choosing extra mates from random families.

It should be noted that the breeder will not be able to use this method of reducing inbreeding if he is applying mass selection for some trait. If the trait is at least partly under genetic control, related individuals will tend to have similar phenotypic values for the trait because of the similarity of their genotypes. Superior phenotypes will be concentrated in certain families, while other families con-

Table 23.1. Effect of variable k on N_e and ΔF

	Case 1	Case 2	Case 3	Case 4
	0	0	0	0
	0	0	2	2
	1	0	2	2
	1	1	2	2
	2	1	2	2
	2	2	2	2
	3	3	2	2
	3	3	2	2
	4	5	2	3
	4	5	4	3
$\Sigma k_i = N$	20	20	20	20
$\bar{k}$	2	2	2	2
σ_k^2	2.0	3.4	0.8	0.6
N_e	19.5	14.4	27.9	30.0
ΔF	0.0256	0.0346	0.0179	0.0167

tain mostly individuals of inferior phenotype. If the breeder chooses the best N offspring, as he must for effective selection, superior families will be overrepresented, inferior families excluded, so that σ_k^2 will be greatly inflated.

For this reason, Falconer (1960, 1981) has suggested within family selection to minimize inbreeding while allowing some selection. The best two individuals from each family are chosen for mating. The attainable selection response is reduced relative to mass selection, because only intrafamily genetic variance can be realized. If, however, inbreeding depression is a serious difficulty in the species involved, or if a very small population must be used, within family selection may be a useful alternative.

In Table 23.1, four cases are presented to illustrate the effect of varying σ_k^2. For each case, we assumed that we started with 10 single pair matings (N = 20) in a diploid population with separate sexes. One mating produced no offspring; the others produced from 4 to 10 offspring each. Family sizes were taken from first litter sizes for 10 single pair matings in an outbred mouse population. From among the offspring we chose 20 mates for the next generation.

In the body of the table, the number of mates chosen from each of the 10 families is shown for each case. Summary statistics are shown below. Because 20 mates were chosen from 10 families in each case, $\bar{k}$ is constant (= 2); but σ_k^2, N_e and ΔF vary.

In Case 1, two families each provided 0, 1, 2, 3 or 4 mates. This gave $\sigma_k^2 = 2$, so that ΔF was 0.0256. In a model population with N = 20, ΔF would be 0.025. The inbreeding rate when $\sigma_k^2 = 2$ was only 2.4% greater than the rate for the model population.

For Case 2, 20 mates were chosen completely at random from the 10 families. Random sampling gave us three families each contributing 0 mates, 2 contributing 1 mate each, etc. The resulting σ_k^2 was 3.4, and ΔF was 0.0346, 38% higher than in the model population.

In Case 3, 2 offspring were chosen from each of the 9 available families. This provided 18 mates; an additional 2 were needed to make up for the infertile mating. These were both chosen from the same fertile family, so that we had 1 family that provided no mates, one that provided 4, and 8 that provided 2 each. σ_k^2 was 0.8, and the inbreeding rate, 0.0179, was 28% less than the rate in the model population.

In Case 4, the same procedure was followed, except that the 2 additional mates were chosen from two different families. Then σ_k^2 was 0.6 and ΔF, 0.0167, 33% less than in the model population.

Had every mating produced at least 2 offspring, it would have been possible to reduce σ_k^2 to 0. The inbreeding rate would then have been 0.0128, 49% lower than the model population inbreeding rate.

During the 1950's, a number of investigators bred "random mating populations" of various species to use as genetic controls (e.g., King et al. 1959). Stipulations for the maintenance of these lines included minimizing σ_k^2 in order to maximize N_e. This would reduce the inbreeding rate as much as possible.

It was also frequently stipulated that matings would be made at random, except that matings between close relatives were to be avoided. Such matings would produce offspring more highly inbred than the average expected on the basis of population size. Their presence would raise the average inbreeding for the whole population. If full sib matings were permitted, for example, in a population of size N, it would be possible for a series of full sib matings to occur in successive generations of one family. This family would become, in effect, a subpopulation of size 2. The inbreeding in the population as a whole would then be a weighted average of the inbreeding for a population of size 2 and a population of size $N-2$, and this rate of inbreeding would be greater than that expected from a population of size N.

Lecture 23 Exercises

1. In a flock of domestic fowl, we have five males and twenty-five females together in a breeding pen. What is ΔF for this flock? (Assume equal fertility of all birds and ignore the correction for absence of selfing.)

2. Recalculate ΔF if (a) one female die, or if (b) one male died.

3. We are selecting two lines of mice, using 5 single pair matings per generation in each. In one line, under mass selection, we obtain 5 mates from one family, 4 from a second, and 1 from a third. No mates come from the other two families. With within family selection in the other line, we obtain 2 mates from each of three families, and 4 mates from the fourth; one mating produced no offspring. What are the ΔF values for these two lines?

Lecture 24 Interaction of Inbreeding with Systematic Forces

The systematic forces, selection, mutation, and migration, tend to drive the frequency of *a* in an infinite population toward some equilibrium value, q_e. Introducing finite population size into the population at equilibrium creates sampling variation around q_e. The effect in a large population may be minor, and easily recognized as sampling error. But if the population is small, sampling variation may completely obscure the very existence of q_e. If, for example, there is strong selection against *aa*, and *a* remains in the infinite population only because of recurrent mutation, q_e will be very small. In generation t of a large line, q_{it} will be some value close to q_e. In a small line, q_{it} may be very large, perhaps even 1; fixation for the allele that is at a selective disadvantage can take place.

Due to sampling variation, q_{it} in generation t of a finite line will differ from q_e for that line. Finite line size forces dispersion, so that $q_{it}-q_e$ will tend to increase with time. At the same time, systematic forces will tend to drive q_{it} back toward q_e. A balance between dispersive and systematic forces may result.

Suppose, for example, that we have both forward and reverse mutation, *A* to *a* at a rate $\mathbf{u}$ and *a* to *A* at a rate $\mathbf{v}$, in a set of lines whose effective size is N. The change in the frequency of *a* alleles in line i, from generation t to generation t+1, caused by mutation, is, from Eq. (15.7),

$$\Delta_{\mathbf{u}} q_{it} = \mathbf{u} p_{it} - \mathbf{v} q_{it} = \mathbf{u} - \mathbf{u} q_{it} - \mathbf{v} q_{it}.$$

Similarly, the change in the average gene frequency over all lines, caused by mutation, is

$$\Delta_{\mathbf{u}} \bar{q}_{it} = \mathbf{u} - \mathbf{u}\bar{q}_{it} - \mathbf{v}\bar{q}_{it}.$$

Then the deviation of the i-th line from the overall average in generation t+1, ignoring for the moment the effect of inbreeding, would be

$$\begin{aligned} & q_{it} + \Delta_{\mathbf{u}} q_{it} - \bar{q}_{it} - \Delta_{\mathbf{u}} \bar{q}_{it} \\ &\quad = q_{it} - \bar{q}_{it} + \mathbf{u} - \mathbf{u} q_{it} - \mathbf{v} q_{it} - \mathbf{u} + \mathbf{u}\bar{q}_{it} + \mathbf{v}\bar{q}_{it} \\ &\quad = q_{it} - \bar{q}_{it} - \mathbf{u}(q_{it} - \bar{q}_{it}) - \mathbf{v}(q_{it} - \bar{q}_{it}) \\ &\quad = (q_{it} - \bar{q}_{it})(1 - \mathbf{u} - \mathbf{v}). \end{aligned}$$

From this it follows that the variance among lines, still ignoring the effects of inbreeding, would be, in generation t+1,

$$V_{t+1}(q_i) = E(q_{it} - \bar{q}_{it})^2 (1 - \mathbf{u} - \mathbf{v})^2 = (1 - \mathbf{u} - \mathbf{v})^2 V_t(q_i).$$

Therefore,

$$\begin{aligned} \Delta_{\mathbf{u}} V_t(q_i) &= V_{t+1} q_i - V_t(q_i) = (1 - \mathbf{u} - \mathbf{v})^2 V_t(q_i) - V_t(q_i) \\ &= -V_t(q_i)[1 - (1 - \mathbf{u} - \mathbf{v})^2] \end{aligned} \qquad (24.1)$$

Mutation tends to decrease variance among the lines; $\Delta V_t(q_i)$ is negative. The decrease in variance can also be deduced from the equation for $V_{t+1}(q_i)$ given above; because $(1-\mathbf{u}-\mathbf{v})^2 < 1$,

$$V_{t+1}(q_i) < V_t(q_i).$$

At the same time, of course, genic drift tends to increase the variance among lines. Due to mutation, $\bar{q}_{it}$ is no longer necessarily equal to q_0; but from Eq. (21.2),

$$V_t(q_i) = (\bar{p}_{it}\bar{q}_{it}/2N) + (1-1/2N)V_{t-1}(q_i).$$

The change in $V_t(q_i)$ due to genic drift is

$$\begin{aligned} \Delta_D V_t(q_i) &= V_{t+1}(q_i) - V_t(q_i) \\ &= [\bar{p}_{it}\bar{q}_{it} - V_{t-1}(q_i)]/2N. \end{aligned} \tag{24.2}$$

Under the combined influences of genic drift and mutation, the net change in variance will be approximately the sum of $\Delta_{\mathbf{u}} V_t(q_i)$ and $\Delta_D V_t(q_i)$. No net change in variance will occur if

$$[\bar{p}_{ie}\bar{q}_{ie} - V_e(q_i)]/2N = [1-(1-\mathbf{u}-\mathbf{v})^2]V_e(q_i).$$

Rearranging this,

$$\begin{aligned} \bar{p}_{ie}\bar{q}_{ie} &= 2N[1-(1-\mathbf{u}-\mathbf{v})^2]V_e(q_i) + V_e(q_i) \\ &= V_e(q_i)[1+2N-2N(1-\mathbf{u}-\mathbf{v})^2]; \end{aligned}$$

$$V_e(q_i)/\bar{p}_{ie}\bar{q}_{ie} = 1/[1+2N-2N(1-\mathbf{u}-\mathbf{v})^2].$$

But

$$(1-\mathbf{u}-\mathbf{v})^2 = 1 + \mathbf{u}^2 + \mathbf{v}^2 - 2\mathbf{u} - 2\mathbf{v} + 2\mathbf{uv};$$

and because $\mathbf{u}$ and $\mathbf{v}$ are very small fractions, $\mathbf{u}^2$, $\mathbf{v}^2$, and $2\mathbf{uv}$ will be negligibly small. We can therefore set

$$(1-\mathbf{u}-\mathbf{v})^2 \cong 1 - 2\mathbf{u} - 2\mathbf{v};$$

and because

$$V_e(q_i)/\bar{p}_{i,e}\bar{q}_{i,e} = F_e,$$

$$\begin{aligned} F_e &= 1/[1+2N-2N(1-2\mathbf{u}-2\mathbf{v})] \\ &= 1/[1+2N-2N+4N(\mathbf{u}+\mathbf{v})] = 1/[1+4N(\mathbf{u}+\mathbf{v})]. \end{aligned} \tag{24.3}$$

As previously noted, Kimura (1955) has pointed out that few lines are fixed, under the model situation, until the q_i have reached an approximately rectangular distribution. This will be achieved when F is approximately 0.33. Therefore, if $F_e < 0.33$ in Eq. (24.3), few lines will have been fixed before variance among the lines has reached equilibrium; or, by analogy in a single line, few loci will become fixed if N is sufficiently large to make $F_e < 0.33$. Setting $F_e = 1/3$, and solving Eq. (24.3) for N, will give us an N value for which equilibrium will be established before any significant amount of fixation has taken place:

$$1/3 > 1/[1+4N(\mathbf{u}+\mathbf{v})];$$

$$3 < 1+4N(\mathbf{u}+\mathbf{v});$$

$$2 < 4N(\mathbf{u}+\mathbf{v});$$

$$N = 1/2(\mathbf{u}+\mathbf{v}). \tag{24.4}$$

If N_e for our lines is greater than the N of Eq. (24.4), most lines in the population (or most loci in a given line) will reach equilibrium before becoming fixed.

If we assume $\mathbf{u}+\mathbf{v}=10^{-5}$, N_e must be greater than 50,000 to prevent significant numbers of lines being fixed. A plant or animal breeder will never be in a position to produce so large a line. If he could do so, an equilibrium between mutation and inbreeding would be of little consequence to him. ΔF would be so small (10^{-5}) that he could ignore it, as he usually ignores mutation.

It has previously been stated that mutation rates for minor genes controlling quantitative traits may be higher than accepted levels for major genes controlling qualitative traits (East 1935, Mukai 1964, Bell 1982). If this is true, the N required to prevent significant fixation might be considerably smaller than 50,000. Kusakabe and Mukai (1984) describe a population in which a balance between mutation and drift appears to maintain a stable polymorphism at intermediate allele frequencies for minor gene loci.

In Equations (17.5) and (17.6), we saw that mutation and migration were comeasurable, both replacing *A* genes with *a*, or vice versa. We can therefore include the effect of migration along with that of mutation in the foregoing argument, and can rewrite Eq. (24.3) to include migration effects:

$$N > 1/2(\mathbf{u}+\mathbf{v}+m). \tag{24.5}$$

Ignoring the negligible effect of mutation, $m > 1/2N$; one migrant gene per generation, or one diploid migrant every second generation, will prevent significant amounts of fixation.

Selection also tends to decrease $V_t(q_i)$, driving q in all lines toward the same end point. If there is positive overdominance, most large lines will reach equilibrium at the q_e dictated by the overdominance. If selection leads to extinction of *a*, most large lines will become fixed for *A*. But if lines are small, *a* may become fixed even in the face of strong selection against it. As in the case of mutation and migration, a value of N can be defined which will demarcate the difference between "large" and "small" lines.

Lecture 24 Exercise

Pursuant to the suggestion that mutation rates for minor genes controlling quantitative traits are higher than those for major genes, suppose that $\mathbf{u}=0.01$, $\mathbf{v}=0.001$, for a minor gene locus. What N would be necessary to prevent significant fixation?

Lecture 25 Coancestry

When we calculate the inbreeding coefficient, F_{it}, for generation t of line i from its effective breeding size, the value which we obtain is the expected inbreeding coefficient for lines of that size in generation t. The average of the individual inbreeding coefficients in a particular line i in generation t will vary around this expected value; and of course, the inbreeding coefficient of individuals within line i in generation t will vary around the line average. The expected value thus does not tell us very much about the actual inbreeding coefficient of any individual. But under some circumstances, it may be useful to know the inbreeding coefficients for specific individuals.

The inbreeding coefficient of an individual A, F_A, is defined as the probability that the two genes of A at a given locus are identical, that is, descended from the same ancestral gene. Wright (1922) used his method of path coefficients to calculate individual inbreeding coefficients. Emik and Terrill (1949) and Cruden (1949) related Wright's individual inbreeding coefficients to Malecot's (1948) coefficient of coancestry. We will define the coefficient of coancestry between individuals A and B, f_{AB}, as the probability that a random gene from A and a random gene from B, at a given locus, are identical. If we can evaluate f_{AB}, we can calculate the inbreeding coefficient of C, the offspring of A × B.

Coancestry coefficients, being essentially algebraic, are more amenable to computer manipulation than are Wright's geometric path coefficients (Hazel and Lush 1950). We will therefore first discuss coancestry calculations (Plum 1954) before turning to a brief consideration of the path method.

Let G(A: X_iX_j) symbolize the probability that the genotype of A is X_iX_j; similarly, G(B: X_kX_l) is the probability that B has genotype is X_kX_l. As before, subscripts on gene symbols will specify genic identities, not allelic forms. The probability of drawing gene X_i at random from individual A, R(X_i, A), is of course, conditional upon the probability that A includes gene X_i in its genotype:

$$R(X_i, A) = R(X_i, A|A: X_iX_j)G(A: X_iX_j).$$

If we define

$$G(A: X_1X_2) = 1 = G(B: X_3X_4),$$

$$R(X_1, A) = (1/2)(1) = 1/2; \; R(X_3, B) = 1/2.$$

The probability of drawing the random pair, X_1 from A and X_3 from B, is

$$R(X_1, A; X_3, B) = R(X_1, A; X_3, B|A: X_1X_2, B: X_3X_4G(A: X_1X_2, B: X_3X_4).$$

But the two drawings are independent, so that

$$R(X_1\,A;\,X_3,B)=R(X_1,A|A{:}\;X_1X_2)G(A{:}\;X_1X_2)R(X_3,B|B{:}\;X_3X_4)$$
$$G(B{:}\;X_3X_4)=(1/2)(1/2)=1/4\,.$$

If now we define $I(X_i, X_k)$ as the probability that genes X_i and X_k are identical, the probability of drawing a pair of genes, one from A and one from B, which are identical, is

$$\begin{aligned} f_{AB}= &\; R(X_1, A; X_3, B)I(X_1, X_3)+R(X_1, A; X_4, B)I(X_1, X_4) \\ &+R(X_2, A; X_3, B)I(X_2, X_3)+R(X_2, A; X_4, B)I(X_2, X_4) \\ =&\;(1/4)[I(X_1, X_3)+I(X_1, X_4)+I(X_2, X_3)+I(X_2, X_4)] \\ =&\;(1/4)\Sigma I(X_i, X_k). \end{aligned} \qquad (25.1)$$

Now consider the coancestry of A with itself. We can evaluate this from Eq. (25.1) if we let B and A be the same individual. X_3 is then the same gene as X_1, X_4 as X_2.

$$f_{AA}=(1/4)[I(X_1, X_1)+I(X_1, X_2)+I(X_2, X_1)+I(X_2, X_2)]$$

But $I(X_i, X_i)=1$, and $I(X_i, X_k)=I(X_k, X_i)$, so that

$$f_{AA}=(1/4)[2+2I(X_1, X_2)]=(1/2)[1+I(X_1, X_2)] \qquad (25.2)$$

$I(X_1, X_2)$ is the probability that A's genotype consists of a pair of identical genes; but that probability is simply the inbreeding coefficient of A, F_A, so that we can write

$$f_{AA}=(1/2)(1+F_A);$$
$$F_A=2f_{AA}-1. \qquad (25.3)$$

Equations (25.3) show the relationship between F_A and f_{AA}; given either of these probabilities, we can evaluate the other. When F_A is at its maximum value of 1, f_{AA} is also 1, which is its maximum value (because f_{AA} is also a probability). On the other hand, when $F_A=0, f_{AA}=1/2$; the coancestry of an individual with itself can never be less than 1/2, because it is always possible to draw the same gene twice, and thus obtain two identical genes, on two random draws from one individual.

Now suppose that A and B mate together to produce an offspring, C. C can have any of four genotypes:

$$G(C{:}\;X_1X_3)=G(C{:}\;X_1X_4)=G(C{:}\;X_2X_3)=G(C{:}\;X_2X_4)=1/4.$$

From Eq. (25.2),

$$f_{CC}=(1/2)[1+I(X_i, X_k)]$$

if C's genotype is X_iX_k, so that

$$f_{CC}=(1/2)\{[1+I(X_1, X_3)]G(C: X_1X_3)+[1+I(X_1, X_4)]G(C: X_1X_4)+....\}$$

$$=(1/8)[4+I(X_1, X_3)+I(X_1, X_4)+I(X_2, X_3)+I(X_2, X_4)]$$

$$=(1/8)[4+\Sigma I(X_i, X_k)].$$

But from Eq. (25.1), $I(X_i, X_k)=4f_{AB}$, so that we can write

$$f_{CC}=(1/2)(1+f_{AB}). \tag{25.4}$$

From Eq. (25.3),

$$f_{CC}=(1/2)(1+F_C);$$

it follows that $F_C=f_{AB}$, that is, that the inbreeding coefficient of C is equal to the coancestry between the parents of C. F_C is the probability that the two genes in C are identical. But the two genes in C are a random choice of one gene from A and one from B, and the probability that a random pair of genes, one from A and one from B, are identical is by definition f_{AB}. If $f_{AB}=0$, $F_C=0$. But for f_{AB} to be zero, A and B, the parents of C, must be unrelated; and if C has unrelated parents, it cannot be inbred. And even with $f_{AB}=F_C=0, f_{CC} = 1/2$. If C is not inbred, the minimum coancestry of C with itself is 1/2.

Now consider the coancestry of C with individual L, f_{CL}. Let

$$G(L: X_5X_6)=1.$$

$$f_{CL}=(1/4)[I(X_1, X_5)+I(X_1, X_6)+I(X_3, X_5)+I(X_3, X_6)]\ G(C: X_1X_3)$$
$$+(1/4)[I(X_1, X_5)+I(X_1, X_6)+I(X_4, X_5)+I(X_4, X_6)]\ G(C: X_1X_4)$$
$$+(1/4)[I(X_2, X_5)+I(X_2, X_6)+I(X_3, X_5)\ +I(X_3, X_6)]G(C: X_2X_3)$$
$$+(1/4)[I(X_2, X_5)+I(X_2, X_6)+I(X_4, X_5)+I(X_4, X_6)]G(C: X_2X_4).$$

In the above equation, each of the possible genotypes for C occurs with probability 1/4, so that the equation can be written as the sum of a series of identity probabilities multiplied by 1/16. Each of the identity probabilities occurs twice; e.g., $I(X_1, X_5)$ occurs if the genotype of C is X_1X_3 and also if it is X_1X_4. Therefore, the above expression can be simplified to

$$f_{CC}=(1/8)[I(X_1, X_5)+I(X_1, X_6)+I(X_2, X_5)+I(X_2, X_6)$$
$$+I(X_3, X_5)+I(X_3, X_6)+I(X_4, X_5)+I(X_4, X_6)].$$

But

$$f_{AL}=(1/4)[I(X_1, X_5)+I(X_1, X_6)+I(X_2, X_5)+I(X_2, X_6)];$$

$$f_{BL}=(1/4)[I(X_3, X_5)+I(X_3, X_6)+I(X_4, X_5)+I(X_4, X_6)].$$

Therefore,

$$f_{CL}=(1/2)(f_{AL}+f_{BL}). \tag{25.5}$$

Equation (25.5) supplies our first averaging rule; the coancestry between C and L is the average of the coancestries between L and the parents of C.

If J and K are the parents of L, from Eq. (25.5),

$$f_{CL}=(1/2)(f_{CJ}+f_{CK})$$
$$=(1/2)[(1/2)(f_{AJ}+f_{BJ})+(1/2)(f_{AK}+f_{BK})]$$
$$=(1/4)(f_{AJ}+f_{BJ}+f_{AK}+f_{BK}) \tag{25.6}$$

Equation (25.6) is our second averaging rule; the coancestry between C and L is the average of the coancestries between the parents of C and the parents of L.

f_{CL} is the probability that a random gene from C is identical with a random gene from L. This probability can be nonzero only if either A or B, the parents of C, carries a gene identical with a gene in either J or K, the parents of L. It is also possible for the two genes in C to be identical with the two genes in L, that is, for C and L to have identical genotypes. For this to happen, however, A must carry a gene identical with a gene in J, and, at the same time, B must carry a gene identical with one in K (or A with K, B with J, simultaneously.) The probability of genotypic identity is therefore a function of the product of coancestry coefficients between the parents of C and L. If either coefficient is 0, that is, if the parents of C and L are unrelated, the product of the coefficients will also be 0. Even if the product is nonzero, it will usually be small.

Lecture 25 Exercise

In the text we have an individual C whose parents are A and B, and an individual L with parents J and K. If A, B, J and K are all unrelated and noninbred, calculate F_C, F_L, and f_{CL}.

Lecture 26 Applications of Coancestry

We can now evaluate the coancestry between specific relatives. Consider first the coancestry between an individual and one of its offspring, f(PO). If A and B are the parents of C, f_{AC} is an example of a parent-offspring coancestry coefficient. From Eq. (25.5),

$$f_{AC}=(1/2)(f_{AA}+f_{AB}).$$

Substituting into this equation the coancestry of A with itself, from Eq. (25.2),

$$f_{AC}=(1/4)(1+F_A+2f_{AB}). \tag{26.1}$$

A parent transmits a copy of one of its genes to each of its offspring. Therefore, f(PO) can never be 0. At least one gene in parent A and one gene in offspring C are identical with probability 1. The probability that it is this gene that is chosen at random from both parent and offspring is 1/4; even if the parent is not inbred and is unrelated to the other parent, $f(\text{PO})=1/4$.

The probability that A's two genes are identical is F_A. If A's genes are identical, the gene transmitted to C must be identical with either. Thus, $F_A > 0$ must increase f_{AC}.

f_{AC} will also be increased if C can receive from B a gene identical with a gene in A. But this can happen only if A and B are related, that is, if $f_{AB} > 0$. Therefore, $f_{AB} > 0$ also increases f_{AC}.

The inbreeding of B does not directly affect f_{AC}, because only one gene from B, the gene transmitted to C, is involved in the coancestry between A and C.

C receives only one of its genes, say X_1, from A. C's other gene, X_3, comes from its other parent, B. Unless X_3 is identical with A's other gene, X_2, A and C cannot have identical genotypes. X_2 in A and X_3 in B can be identical only if A and B are related, that is, if $f_{AB} > 0$. Unless A and B are related, the probability that A and C have identical genotypes is 0.

Half sibs are individuals with one parent in common. If A produces offspring C from its mating with B, and offspring E from a second mating with D, C and E are half sibs, with common parent A. The coancestry between half sibs, f(HS), is, from Eq. (25.6),

$$\begin{aligned} f_{CE}&=(1/4)(f_{AA}+f_{AB}+f_{AD}+f_{BD})\\ &=(1/8)(1+F_A+2f_{AB}+2f_{AD}+2f_{BD}). \end{aligned} \tag{26.2}$$

Individual A transmits a copy of one of its genes to C, and also a copy of one of its genes to E. The probability that both are copies of the same gene in A is 1/2, while the probability that the gene from A is then sampled from both C and E is 1/4. Therefore, $f_{CE}= 1/8$, even if A is not inbred, and A, B and D are all unrelated; 1/8 is the minimum value of f(HS).

Inbreeding of A obviously increases the probability that C and E receive identical genes from A, and therefore increases f_{CE}. Aside from copies of the same or identical genes in A, C and E can have identical genes if

1. B transmits to C a gene identical with the gene that A transmits to E ($f_{AB} > 0$);
2. D transmits to E a gene identical with the gene that A transmits to C ($f_{AD} > 0$);
or
3. B transmits to C a gene identical with the gene that D transmits to E ($f_{BD} > 0$).

Half sibs can have identical genotypes if they receive copies of the same gene from their common parent, and at the same time receive identical genes from the parent of each not common to the other. This can happen only if the parents not in common are related; if B and D are unrelated, the probability that half sibs C and E have identical genotypes is 0.

Finally, full sibs have both parents in common. If A and B produce two offspring, C and M, the latter two are full sibs, and their coancestry is f(FS). From Eq. (25.6).

$$\begin{aligned} f_{CM} &= (1/4)(f_{AA} + f_{AB} + f_{AB} + f_{BB}) \\ &= (1/8)(2 + F_A + F_B + 2f_{AB}) \\ &= (1/4)[1 + (F_A/2) + (F_B/2) + f_{AB})]. \end{aligned} \tag{26.4}$$

C and M can be regarded as half sibs through common parent A and simultaneously half sibs through common parent B. Even if A and B are unrelated and neither is inbred, f_{CM} will be 1/4. The minimum value of f(FS) is therefore 1/4. Inbreeding of either parent, or a relationship between the two parents, will increase f(FS).

A pair of full sibs, C and M, can both receive copies of the same gene from parent A and also from parent B, even if A and B are unrelated and have no common identical genes. Thus, full sibs can have identical genotypes even if their parents are not related.

We have summarized the minimum values of f for the types of relatives discussed above in Table 26.1. By minimum values, we refer to the values of these functions when all parents are unrelated and noninbred.

Table 26.1. Minimum values of coancestries for specific relatives (assuming parents unrelated and noninbred)

Relatives	f
Parent-offspring	0.25
Half sibs	0.125
Full sibs	0.25

It must be emphasized that the genic coancestry of an individual with itself cannot be obtained by applying the averaging rules. The coancestry of C with itself, from Eq. (25.4), is

$$f_{CC}=(1/2)(1+f_{AB}).$$

If we attempt to apply the first averaging rule, Eq. (25.5),

$$(1/2)(f_{AC}+f_{BC})=(1/8)(2+F_A+F_B+2f_{AB});$$

this is the genic coancestry of C, not with itself, but with a full sib. Similarly, the second averaging rule gives the full sib coancestry if we attempt to use it to evaluate f_{CC}. The averaging rules give the coancestries between individuals with the same parents, but not of an individual with itself. The latter must be calculated from Eq. (25.4).

Lecture 26 Exercises

1. Suppose that A and J are full sibs, offspring of unrelated and noninbred parents. Calculate f_{CL}, where C is the offspring of A x B and L that of J x K. Assume B and K are noninbred and unrelated to each other or to A or J.

2. Recalculate f_{CL} assuming that B and K are also full sibs from a mating between unrelated and noninbred parents. Assume that B and K are unrelated to A and J.

Lecture 27 Tabular Calculation of Coancestry

If L is the offspring of J × K, and if we know the coancestries of A and B with J and K, we can calculate the coancestries of A and B with L, using the first averaging rule, Eq. (25.5). These calculations lend themselves readily to a tabular format:

	J	K	L
A	f_{AJ}	f_{AK}	$f_{AL} = (f_{AJ} + f_{AK})/2$
B	f_{BJ}	f_{BK}	$f_{BL} = (f_{BJ} + f_{BK})/2$

The entry in row A, column L, is the average of the entries in row A, columns J and K; the row B entries in column L are similarly the average of the entries in columns J and K. Emik and Terrill (1949), Cruden (1949), Hazel and Lush (1950) and Plum (1954) all emphasized the simplicity of calculating coancestry coefficients using a tabular format.

We can expand this concept to construct an n x n table which will give all of the pairwise coancestries among n individuals in a pedigree. Consider, for example, the pedigree given in Fig. 27.1. We wish to calculate the inbreeding coefficient of individual N in generation 6 of this pedigree. We will assume that the

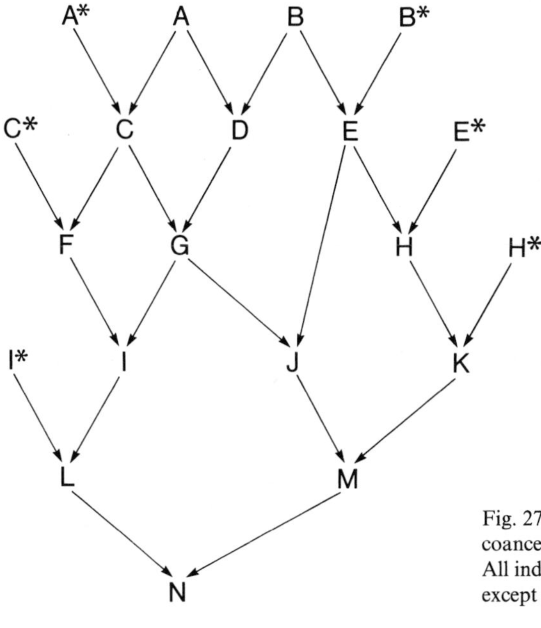

Fig. 27.1. A pedigree for tabular calculation of coancestry coefficients (after Falconer, 1960). All individuals are noninbred and unrelated except as shown in the pedigree

four individuals in generation 1, A*, A, B, and B*, are noninbred and unrelated, and also that individuals C* and E* in generation 2, H* in generation 3, and I* in generation 4 are noninbred and unrelated to any other individuals except as shown by the pedigree.

Table 27.1. Generation 1 coancestries

	A*	A	B	B*
A*	0.5	0	0	0
A	0	0.5	0	0
B	0	0	0.5	0
B*	0	0	0	0.5

Table 27.2. Coancestries between generation 1 and 2

	A*	A	B	B*	C*	A* × A C	A × B D	B × B* E	E*
A*	0.5	0	0	0	0	0.25	0	0	0
A	0	0.5	0	0	0	0.25	0.25	0	0
B	0	0	0.5	0	0	0	0.25	0.25	0
B*	0	0	0	0.5	0	0	0	0.25	0

Table 27.4. Complete table of coancestries for the pedigree in Fig. 27

	A*	A	B	B*	C*	A* × A C	A × B D	B × B* E	E*
A*	0.5000	0	0	0	0	0.2500	0	0	0
A	0	0.5000	0	0	0	0.2500	0.2500	0	0
B	0	0	0.5000	0	0	0	0.2500	0.2500	0
B*	0	0	0	0.5000	0	0	0	0.2500	0
C*	0	0	0	0	0.5000	0	0	0	0
C	0.2500	0.2500	0	0	0	0.5000	0.1250	0	0
D	0	0.2500	0.2500	0	0	0.1250	0.5000	0.1250	0
E	0	0	0.2500	0.2500	0	0	0.1250	0.5000	0
E*	0	0	0	0	0	0	0	0	0
F	0.1250	0.1250	0	0	0.2500	0.2500	0.0625	0	0
G	0.1250	0.2500	0	0	0	0.3125	0.3125	0.0625	0
H	0	0	0.1250	0.1250	0	0	0.0625	0.2500	0
H*	0	0	0	0	0	0	0	0	0
I*	0	0	0	0	0	0	0	0	0
I	0.1250	0.1875	0	0	0.1250	0.2812	0.1875	0.0312	0
J	0.0625	0.1250	0.1250	0.1250	0	0.1562	0.2188	0.2812	0
K	0	0	0.0625	0.0625	0	0	0.0312	0.1250	0.1250
L	0.0625	0.0938	0	0	0.0625	0.1406	0.0928	0.0156	0
M	0.0312	0.0625	0.0625	0.0625	0	0.0781	0.1250	0.1468	0.0625
N	0.0468	0.0782	0.0312	0.0312	0.0312	0.0703	0.1094	0.0812	0.0312

$F_N = 2(f_{NN}) - 1 = f_{LM} = 0.0508$

We first construct a table showing the coancestries among the individuals in generation 1. In this 4 x 4 table (Table 27.1), each leading diagonal entry is the coancestry of an individual with itself. Because we assume these individuals to be noninbred, each of the leading diagonal entries is 0.5, the coancestry of a noninbred individual with itself. The off-diagonal entries are coancestries between pairs of these four individuals, 0 in each case because they are assumed to be unrelated.

Next we add columns to the right of the above table for generation 2 individuals, C*, C, D, E, and E* (Table 27.2). C* and E* are unrelated to any generation

Table 27.3. Coancestries within generation 2

	A*	A	B	B*	C*	A* × A C	A × B D	B × B* E	E*
A*	0.5	0	0	0	0	0.25	0	0	0
A	0	0.5	0	0	0	0.25	0.25	0	0
B	0	0	0.5	0	0	0	0.25	0.25	0
B*	0	0	0	0.5	0	0	0	0.25	0
C*	0	0	0	0	0.5	0	0	0	0
C	0.25	0.25	0	0	0	0.5	0.125	0	0
D	0	0.25	0.25	0	0	0.125	0.5	0.125	0
E	0	0	0.25	0.25	0	0	0.125	0.5	0
E*	0	0	0	0	0	0	0	0	0.5

C* × C F	C × D G	E × E* H	H*	I*	F × G I	G × E J	H × H* K	I* × I L	J × K M	L × M N
0.1250	0.1250	0	0	0	0.1250	0.0625	0	0.0625	0.0312	0.0468
0.1250	0.2500	0	0	0	0.1875	0.1250	0	0.0938	0.0625	0.0782
0	0.1250	0.1250	0	0	0.0625	0.1875	0.0625	0.0312	0.0938	0.0625
0	0	0.1250	0	0	0	0.1250	0.0625	0	0.0625	0.0312
0.2500	0	0	0	0	0.1250	0	0	0.0625	0	0.0312
0.2500	0.3125	0	0	0	0.2812	0.1562	0	0.1406	0.0781	0.0703
0.0625	0.3125	0.0625	0	0	0.1875	0.2188	0.0312	0.0938	0.1250	0.1094
0	0.0625	0.2500	0	0	0.0312	0.2812	0,1250	0.0156	0.1468	0.0812
0	0	0.2500	0	0	0	0	0.1250	0	0.0625	0.0312
0.5000	0.1562	0	0	0	0.2781	0.0781	0	0.1390	0.0390	0.0890
0.1562	0.5625	0.0312	0	0	0.3594	0.3125	0.0156	0.1797	0.1640	0.1718
0	0.0312	0.5000	0	0	0.0156	0.1406	0.2500	0.0878	0.1953	0.1016
0	0	0	0.5000	0	0	0	0.2500	0	0.1250	0.0625
0	0	0	0	0.5000	0	0	0	0.2500	0	0.1250
0.2781	0.3594	0.0156	0	0	0.5781	0.1953	0.0078	0.2890	0.1015	0.1953
0.0781	0.3125	0.1406	0	0	0.1953	0.5312	0.0703	0.0976	0.3008	0.1992
0	0.0156	0.2500	0.2500	0	0.0078	0.0703	0.5000	0.0039	0.2852	0.1446
0.1390	0.1797	0.0078	0	0.2500	0.2890	0.0976	0.0039	0.5000	0.0508	0.2754
0.0390	0.1640	0.1953	0.1250	0	0.1015	0.3008	0.2852	0.0508	0.5352	0.2930
0.0890	0.1718	0.1016	0.0625	0.1250	0.1953	0.1992	0.1446	0.2754	0,2930	0.5254

1 individual, so that all the entries in these two columns are 0. But C is the offspring of A* × A; to remind ourselves of that fact, we write "A* × A" above C at the top of that column. Therefore, the entry in column C of a given row is the average of the entries for A* and A in that same row. Similarly, D and E are the offspring of A × B and B × B*, respectively. The entries in column D are therefore averages of the entries in columns A and B, and column E entries are averages of entries in columns B and B*, of the same row.

Row C, column A* is f_{CA^*}; row A*, column C, f_{A^*C}. As these two coancestries are the same, the corresponding entries will be the same. We can copy a third section of the table, rows C* to E*, columns A* to B*, from the section just completed, rows A* to B*, columns C* to E* (Table 27.3); the completed table will be symmetrical about its leading diagonal.

This leaves us with the section of Table 27.3 representing the coancestries within generation 2 (rows C* to E*, columns C* to E*). The leading diagonal of this section is again the coancestries of individuals with themselves. From Eq. (25.4), these coancestries are

$$f_{SS} = (1/2)(1 + f_{PP*}),$$

where P and P* are the parents of S. Thus, the entry in row C, column C will be 0.5 plus half of the entry in row A*, column A (or row A, column A*). The leading diagonal entries for all generation 2 individuals are 0.5, because their generation 1 parents have all been assumed to be unrelated. Therefore, individuals C, D and E are noninbred because they have unrelated parents, while C* and E* have been assumed to be noninbred.

The off-diagonal entries in the section representing coancestries among generation 2 individuals are again calculated by averaging entries in the columns representing an individual's parents. C* and E* are assumed to be unrelated to any generation 1 or 2 individuals, and therefore all off-diagonal entries in these columns are 0. These entries complete Table 27.3.

We continue in this manner until we have developed the complete coancestry table for the whole pedigree of 20 individuals (Table 27.4). To calculate the inbreeding coefficient of individual N, we use N's self-coancestry:

$$F_N = 2(f_{NN}) - 1 = 2(0.5254) - 1 = 0.0508.$$

Alternatively, we can look up the entry in row L, column M of the table, which represents the coancestry between the parents of N:

$$F_N = f_{LM} = 0.0508.$$

This procedure, though somewhat tedious, can be applied to hand calculation of coancestries for any pedigree. In addition, it readily lends itself to computer manipulation. With an appropriate program, the whole table can be produced by specifying the parentage of each individual in the pedigree. Various shortcuts are possible. If, for example, every generation is discrete (in our example, they were not) it would only be necessary to calculate coancestries among individuals within each generation. The coancestries within one generation could then be used to calculate those within the next, using the second averaging rule. This would be somewhat difficult if the calculation is being done by hand, but is easily possible on the computer. By hand, with discrete generations, a series of tables for successive pairs of generations, like Table 27.3, can be developed; the coancestries within generation 2 from Table 27.3 could then be used as a starting point for the next table showing coancestries between generations 2 and 3, and so forth.

Lecture 27 Exercises

1. Recalculate the coancestry of a pair of double first cousins (Exercise 2, Lecture 26), using the tabular format.

2. Fig. 27.2 shows another pedigree. Calculate the coancestries for this pedigree.

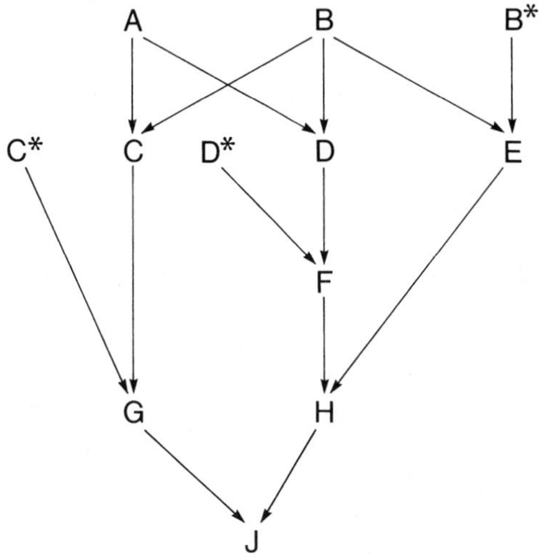

Fig. 27.2. Another sample pedigree. All individuals are noninbred and unrelated except as shown in the pedigree

Lecture 28 Path Calculation of F

Wright (1922) originally derived his system for calculating individual inbreeding coefficients from his general method of path coefficients, or standardized partial regression coefficients (Wright 1934). As used in calculating inbreeding, paths point out the flow of identical genes, and the coefficients are identity probabilities. The path coefficient method constitutes a geometric approach to analysis, as opposed to the algebraic approach of Fisherian statistics. Path coefficients are still employed for many analytical purposes, especially in human genetics (Li 1975) and the social sciences, although somewhat out of favor in quantitative genetics.

The inbreeding coefficient is the probability that an individual receives identical genes from its two parents. Obviously, this probability is 0 unless the parents have one or more ancestors in common. Consider the pedigree shown in Fig. 28.1. We wish to calculate F_J, the inbreeding coefficient of individual J. G and H, the parents of J, have a common ancestor, A; we assume that the members of the pedigree are unrelated except as shown by the pedigree. F_J, then, is the probability that J receives simultaneously from G and H identical genes originating in A.

Let A's genotype be symbolized as $X_A X_{A*}$. The probability that A transmits X_A to C is then 1/2. If this occurs, the probability that C transmits X_A to its offspring G is also 1/2; and the probability that, in that event, G transmits X_A to J, 1/2. The probability that J receives X_A from A via C and G is the joint probability of all three of these events, $(1/2)^3$.

The three steps, from A to C, from C to G, and from G to J, form a path. The coefficient associated with each step in the path is the probability that X_A passes through that step, and the coefficient of the whole path is the product of the individual coefficients for each step.

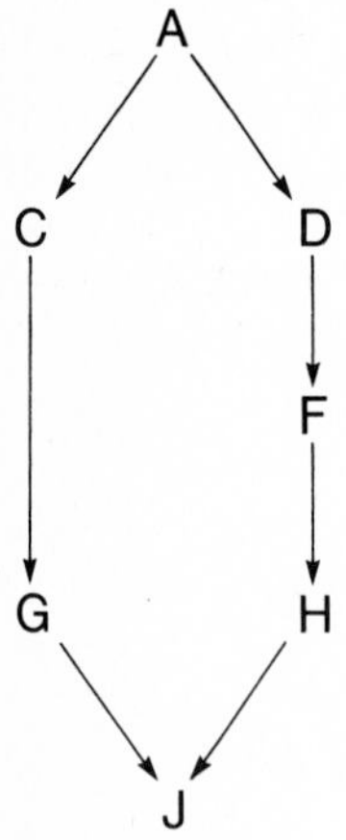

Fig. 28.1. Pedigree for the path calculation of inbreeding (Falconer 1960). All individuals can be assumed to be noninbred and unrelated except as shown in the pedigree

It is also possible for X_A to pass from A to J by way of D, F and H. This is a path of four steps, each with coefficient 1/2, so that the probability that J receives X_A from A via D, F, H is $(1/2)^4$.

If these two paths are realized simultaneously, J will have genotype $X_A X_A$. The probability that both paths will be realized is the product of their separate probabilities:

$$G(J: X_A X_A) = (1/2)^3 (1/2)^4 = (1/2)^{3+4}.$$

Individual A can also transmit gene X_{A^*} to J via the same two paths, so that

$$G(J: X_{A*} X_{A*}) = (1/2)^{3+4}.$$

Therefore, the probability that J receives two copies of the same gene in A, via C, G and via D, F, H is

$$G(J: X_A X_A \text{ or } X_{A*} X_{A*}) = 2(1/2)^{3+4} = (1/2)^{3+4-1}.$$

J can receive X_A from A via C, G, and at the same time receive X_{A*} from A via D, F, H. The probability of this event is again $(1/2)^{3+4}$. Because J can also receive X_{A*} via C, G and X_A via C, D, F with the same probability,

$$G(J: X_A X_{A*}) = (1/2)^{3+4-1}.$$

But X_A and X_{A*} can be identical, with probability F_A. Therefore, the probability that J receives identical genes X_A and X_{A*} from A is

$$G(J: X_A X_{A*}) I(X_A, X_{A*}) = (1/2)^{3+4-1} F_A.$$

Finally, therefore, the probability that I receives identical genes from A is the inbreeding coefficient of J,

$$F_J = (1+F_A)(1/2)^{3+4-1}.$$

[The sum

$$G(J: X_A X_A) + G(J: X_{A*} X_{A*}) + G(J: X_A X_{A*}) = 2(1/2)^6 = 1/32$$

is less than 1. It is not a certain event that J will receive a gene from A by either path. For example, C might transmit to G the gene it received from its other parent, rather than the one it received from A. Then J cannot receive a gene from A via the path A-C-G-J. Every individual in either path can similarly fail to transmit an A gene.]

In general, if the two parents of an individual have a common ancestor, the probability that the individual will receive identical genes originating in that ancestor from both parents will be

$$F_J = (1+F_A)(1/2)^{n+m-1}, \tag{28.1}$$

where J is the individual, A the ancestor common to both of J's parents, and the paths from A to J have n steps going through one of J's parents and m steps going through the other.

The pedigree in Fig. 28.1 is a relatively simple one. It is, however, only part of a more complex pedigree, which we have already presented in Fig. 27.2. It is

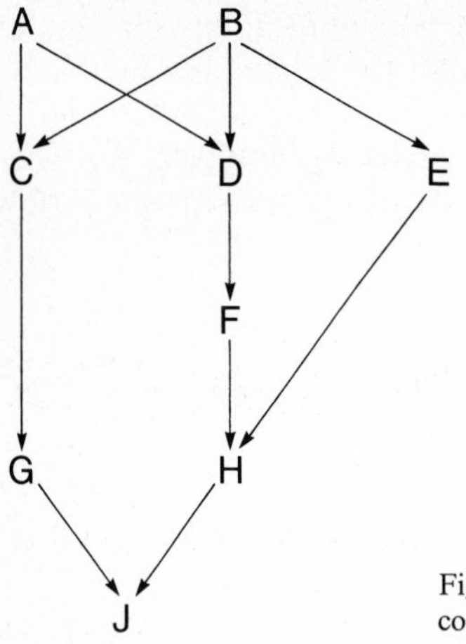

Fig. 28.2. An expanded pedigree for path calculation of inbreeding coefficients (equivalent to Fig. 27.2)

presented again in Fig. 28.2, with individuals B*, C*, and D*, who have no effect upon the path evaluation of F_J, eliminated.

This figure shows that, in addition to the common ancestor A as in Fig. 28.1, G and H have a second common ancestor, B. Also, H is descended from B by two chains of descent, through D and F, and also through E. Because J can receive identical genes from B, and because B genes can descend to H through either of the above paths, F_J must include all paths through which G and H can receive identical genes from any common ancestor and transmit them to J:

$$F_J = \Sigma(1+F_i)(1/2)^{n(i)+m(i)-1}.$$

In the pedigree of Fig. 28.2, the paths are

J−G−C−A−D−F−H−J; n(i) = 3, m(i) = 4;

J−G−C−B−D−F−H−J; n(i) = 3, m(i) = 4;

and J−G−C−B−E−H−J; n(i) = 3, m(i) = 4.

Thus,

$$F_J = (1+F_A)(1/2)^6 + (1+F_B)(1/2)^6 + (1+F_B)(1/2)^5.$$

If we assume that A and B are noninbred,

$$F_J = (1/2)^6 + (1/2)^6 + (1/2)^5 = (1/2)^4 = 0.0625.$$

This result agrees with the calculation made in Lecture 27, Exercise 2, using coancestry methods.

Lecture 28 Exercise

Calculate the inbreeding coefficient of N in the pedigree given in Fig. 27.1, using path methods. Assume common ancestors are not inbred except as given in the pedigree.

Lecture 29 Systematic Inbreeding

Intentional, systematic, rapid inbreeding has been useful, both in genetic research and in plant and animal breeding. Mendel's success in deciphering the action of his "unit factors" in pea plants was in part attributable to the fact that these plants are naturally self fertilizing. Having made a cross between lines with opposing characteristics, he could then allow inbreeding to sort out the results over the next several generations.

Clarence Cook Little, at Harvard's Bussey Institute in the early 1900's, developed inbred mouse strains by full sib mating over many generations (Little 1916). Within such a strain, tissues, tumor tissues in particular, could be grafted from one mouse to another, and thus kept alive for study, long before in vitro tissue culture techniques were developed (Tyzzer and Little 1916). Within an inbred strain, also, genotypes at a mutant locus could be compared on a uniform genetic background, to assess and characterize more accurately effects of a mutant gene.

The early corn breeders, Shull, East, Jones, and others (e.g., East 1908), self-fertilized maize plants for a series of generations to develop inbred lines bearing certain advantageous characteristics, then crossed these lines to produce high yielding hybrid corn. Similar procedures have been used in other crop plants and, with less success, in animals.

Crossing inbred lines is necessary because inbreeding tends to have a depressing effect. Reproduction, viability, size, and luxuriance all are likely to be less in an inbred than in the outbred line from which it was developed. Crossing inbreds reverses the depression; indeed, the hybrid frequently exceeds not only its inbred parents, but the foundation stocks from which the inbreds originated, with regard to these traits. The cause of these antagonistic phenomena, inbreeding depression and hybrid vigor, have been the subject of much discussion.

The explanation of hybrid vigor and inbreeding depression by dominance, first put forward by Bruce (1910) and by Keeble and Pellew (1910), argues that these phenomena arise from the tendency for deleterious genes to be recessive. Inbreeding increases the frequency of homozygotes in which the deleterious effects of recessive genes are not masked by an accompanying dominant allele. Since this would be true at each of a number of such loci, the frequency of individuals bearing multiple deleterious genotypes would likewise be increased, as pointed out by Collins (1921). Crossing, especially if the lines crossed are from unrelated foundation stocks, increases heterozygosity and leads to the masking of deleterious recessive genes.

East (1908), Shull (1911, 1914), and Jones (1918) argued that an interaction between opposite alleles in a heterozygote, that is, overdominance, was the true explanation of hybrid vigor. To these workers, inbreeding depression was simply the loss of overdominant effects due to the decreasing frequency of heterozygotes.

Collins (1921) pointed out that in practice it would be difficult to distinguish between the dominance and overdominance theories of hybrid vigor and inbreeding depression, although he felt that dominance was an adequate explanation of these phenomena as observed in corn. Crow (1948) argued that inbreeding depression could usually be explained by the dominance theory, but that hybrid vigor often exceeded the probable limits of the dominance effect. Overdominance appeared to be the best candidate to explain this excess. More recently, Crow (1986) has indicated that, because mutation rates may be higher and partial dominance more prevalent than was realized in 1948, the argument for overdominance has been weakened. Except in specific cases, dominance is probably an adequate explanation for inbreeding depression and hybrid vigor.

Whatever the explanation, inbreeding depression undoubtedly does occur. Nevertheless, inbreeding has been and will continue to be a useful tool for both the geneticist and the breeder. We will therefore look at the effects of several systematic inbreeding methods.

Let us first consider the most rapid form of inbreeding, selfing. Selfing is not possible, of course, in species with separate sexes. But it is possible in many plant species, and in some it may be the normal form of reproduction.

Under selfing, $N=N_e=1$, and $\Delta F=0.5$, so that we expect the inbreeding coefficient to increase rapidly. There is only one individual, A_t, in generation t of a line being selfed (see Fig. 29.1). The average inbreeding coefficient in generation t, F_t, is therefore simply A_t's inbreeding coefficient. From Eq. (25.4), F_t will therefore be the coancestry between A_t's parents. But A_{t-1} is both parents of A_t, so to speak, and

$$F_t = f_{(S)t-1},$$

where $f_{(S)t-1}$ is the coancestry of A_{t-1} with itself. From Eq. (25.3),

$$f_{(S)t} = (1/2)(1+F_{t-1});$$

therefore,

$$F_t = (1/2)(1+F_{t-1}). \qquad (29.1)$$

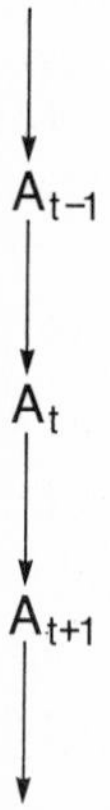

Fig. 29.1. Pedigree for inbreeding by selfing. There is only one individual, A_t, in generation t

It is somewhat easier to calculate successive values of the inbreeding coefficient from successive values of the panmictic index. Substituting $P_t = 1 - F_t$ into Eq. (29.1),

$$P_t = 1 - (1/2)(1 + 1 - P_{t-1})$$
$$= (2 - 2 + P_{t-1})/2 = P_{t-1}/2.$$

The panmictic index is halved in each generation. Beginning with a noninbred individual ($P_0 = 1$, $F_0 = 0$), successive values of P_t will be 1, 0.5, 0.25, 0.125, etc., and of F_t, 0, 0.5, 0.75, 0.875, etc. F_t will exceed 0.95 by generation 5, 0.99 by generation 7, and 0.999 by generation 10.

The closest inbreeding that we can manage in a species with separate sexes is full sib mating, mating pairs of offspring from the same parents. Then $N = 2$, $N_e = 2.5$ [Eq. (23.1)], and $\Delta F_{FS} = 0.2$, compared to $\Delta F_S = 0.5$ under selfing.

Generation t of a line under full sib mating consists of a pair of sibs, A_t and B_t (see Fig. 29.2). Both are the offspring of mating $A_{t-1} \times B_{t-1}$, so both will have the same inbreeding coefficient, F_t. From Eq. (25.3), that inbreeding coefficient will be the coancestry between A_{t-1} and B_{t-1},

$$F_t = f_{t-1}.$$

From Eq. (25.6), f_{t-1} will be the average of the coancestry of A_{t-2} with itself, B_{t-2} with itself, and twice the coancestry between A_{t-2} and B_{t-2}. But A_{t-2} and B_{t-2} have the same self coancestries, $f_{(S)t-2}$. Therefore,

$$f_{t-1} = (1/4)(2f_{(S)t-2} + 2f_{t-2})$$
$$= (1/2)(f_{(S)t-2} + f_{t-2}).$$

Now, from Eq. (25.3),

$$f_{(S)t-2} = (1/2)(1 + F_{t-2});$$

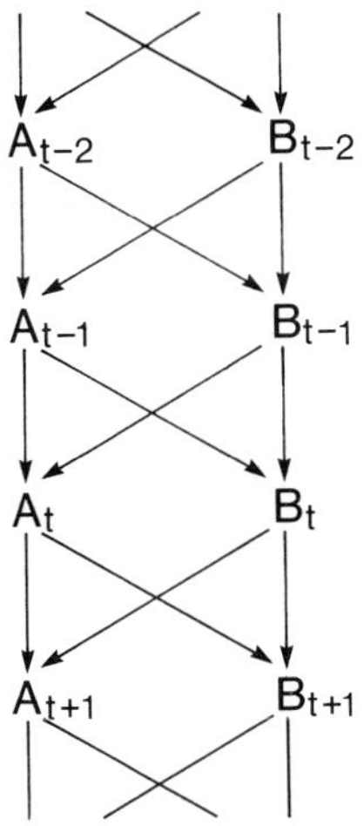

Fig. 29.2. Pedigree for inbreeding by full sib mating. A pair of full sibs, A_t and B_t, are mated together in generation t

and from Eq. (25.4),

$$f_{t-2} = F_{t-1}.$$

Consequently,

$$F_t = f_{t-1} = (1/2)[(1/2)(1 + F_{t-2}) + F_{t-1}]$$
$$= (1/4)(1 + F_{t-2} + 2F_{t-1}). \qquad (29.2)$$

Again, computations are simpler in terms of the panmictic index:

$$P_t = 1 - (1/4)[1 + (1 - P_{t-2}) + 2(1 - P_{t-1})]$$
$$= (1/4)(P_{t-2} + 2P_{t-1}) = (P_{t-1}/2) + (P_{t-2}/4).$$

Starting with $P_0 = 1$, successive P values will be 1, 0.75, 0.625, 0.5, etc., and successive values of F will be 0, 0.25, 0.375, 0.5, etc. It takes more than twice as long for F_t to reach a given value as under selfing; 14 generations to reach 0.95, 20 to reach 0.99.

The third type of systematic inbreeding we will discuss is repeated backcrossing. A backcross is a mating between an individual and his lineal descendant: sire x daughter, granddam x grandson, etc. A repeated backcross mates, say, sire x daughter, then sire x daughter's daughter, etc. This type of breeding has been used by race horse breeders, and by breeders of show animals of various species, to concentrate the genes of a superior individual.

In a repeated backcross we have in generation t an individual, B_t, mated to its parent A; A is also a parent of B_{t-1}, the other parent of B_t (see Fig. 29.3). The inbreeding coefficient of B_t is F_t, and

$$F_t = f_{A,t-1},$$

where $f_{A,t-1}$ is the coancestry between A and B_{t-1}. From Eq. (25.5),

$$f_{A,t-1} = (1/2)(f_{AA} + f_{A,t-2}).$$

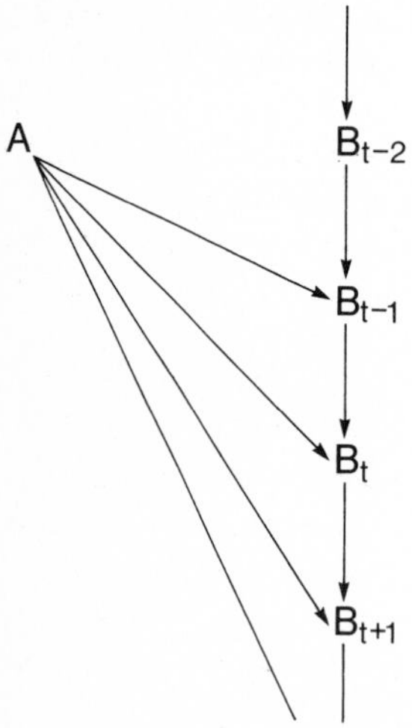

Fig. 29.3. Pedigree for inbreeding by repeated backcrossing. In generation t, individual B_t is backcrossed to its parent (grandparent, greatgrandparent, ...), A

The coancestry of A with itself is

$$f_{AA} = (1/2)(1 + F_A),$$

and this is constant in every generation. Therefore,

$$F_t = (1/4)(1 + F_A + 2F_{t-1}); \tag{29.3}$$

this equation is similar in form to that for the inbreeding coefficient for full sibs, with F_A taking the place of F_{t-2}. The panmictic index form is thus also similar to the full sib equation:

$$P_t = (P_{t-1}/2) + (P_A/4).$$

If A and the original B are both noninbred, the progression of P_t values is 1, 0.75, 0.625, 0.5625, etc. The limiting value of P in this progression is not 0 but 1/2; if $P_{t-1} = 1/2$, then

$$P_t = (1/2)(1/2) + 1/4 = 1/2.$$

The general term for P_t is

$$P_t = P_0/2^t + (2^t - 1)P_A/2^{t+1}.$$

As t becomes very large, $P_0/2^t$ goes to 0; $(2^t - 1)P_A/2^{t+1}$ goes to $P_A/2$. Thus, the minimum P value is $P_A/2$, which will be 1/2 if A is not inbred.

If A is completely inbred, however, Eq. (29.3) becomes

$$F_t = (1/2)(1 + F_{t-1}),$$

which is the same as Eq. (29.1), the inbreeding coefficient in generation t under selfing. The limiting value of P_t is then 0.

True repeated backcrossing to the same individual is obviously limited in its applicability. The original A individual will be available for mating to his descendants for only a few generations. Various adaptations of the scheme, collectively referred to as line breeding, have been employed, such as parent x offspring matings in each generation. We will not analyze these systems in detail.

In one set of circumstances, however, the equivalent of repeated backcrossing to the same individual can be maintained over a long series of generations. If individual A is a member of a highly inbred strain, the original B individual coming from outside that strain, descendants of the A x B cross can be mated in each generation to successive generations of inbred strain members. Since all members of the inbred strain, in whatever generation, have virtually identical genotypes, this system is equivalent to repeated backcrossing to a single individual, that individual having $F_A = 1$.

This procedure is widely used in laboratory mouse breeding to place a mutant gene on an inbred background. To define the effect of a mutation, it is often convenient to place it in a standard homogeneous genotype. Suppose that mutant gene *m* has arisen in a heterogeneous stock of mice, and we wish to assess its effects. This can be done to best effect if it is in a homogeneous genotype, where we can compare *MM*, *Mm*, and *mm* genotypes without the extraneous effects of other differences in genotype. We therefore want to transfer the mutant gene into inbred strain A. We cross heterogeneous carriers of *m* with members of strain A.

Carriers of *m* among the F_1 offspring of this cross are backcrossed to strain A mice, their carrier offspring again backcrossed, and so forth. The eventual result will be a line identical to strain A in genotype except for the substitution of gene *m* for its allele, *M*, originally present in strain A.

The mutant gene *m* will carry with it a small segment of chromosome from the original heterogeneous stock. The line resulting from repeated backcrossing will therefore differ from strain A, not only at the M locus, but at a few loci on either side of it as well. The length in crossover units of the segment of heterogeneous chromosome remaining after t generations can be estimated (Green and Doolittle 1963).

In practice, the simple process outlined above can become complicated. If *mm* homozygotes are crossed to the inbred strain, their progeny are *Mm* heterozygotes. Backcrossing these to the inbred will produce *MM* and *Mm* backcross progeny; if *m* is recessive, it is not possible to identify the heterozygous carriers of *m*, needed for the next backcross, by their phenotypes. It is then necessary to mate the backcross progeny inter se; these matings will produce *mm* offspring when two heterozygotes are mated together. If the mutation has no severe deleterious effects, homozygous *mm* progeny from the inter se matings can be backcrossed to the inbred, starting the cycle over again. But if *mm* is deleterious, it may be necessary to use *Mm* parents, identified by the inter se matings, to backcross to the inbred. Often it may be convenient to create an inbred line which segregates for *M* and *m*, rather than one homozygous for *mm*; again, *Mm* mice must be used in backcrossing at least in the terminal generation. Various systems alternating backcross and inter se generations have been suggested; Green and Doolittle (1963) reviewed the rate of approach to homozygosity under several of these systems.

Despite these complications, backcrossing to an inbred strain approaches complete inbreeding as rapidly as selfing; nine backcross generations are usually accepted as making the line inbred, as compared to 20 generations of full sib mating.

Lecture 29 Exercise

We want to place the head spot gene, *hs*, which occurs in a heterogeneous stock of mice, on an inbred background. Our alternatives are repeated backcrossing to a standard inbred line, C57BL/6, or full sib mating among carriers of *hs*. We want to derive an inbred line in which we can compare *HsHs*, *Hshs* and *hshs* genotypes. Because *hs* is recessive to *Hs*, we must identify carriers in each generation. What are some of the considerations we must keep in mind in deciding between backcrossing and full sib mating among carriers?

Lecture 30 Assortative Mating

One of the most important conditions under which we originally derived the Hardy-Weinberg law was that of random mating. To produce an offspring, a pair of mates was chosen at random. The probabilities of choosing a given genotype for either of the two mates were independent, and were each equal to the frequency of that genotype in the population. Then, if A was the first mate chosen and B the second, in a population where D_t was the generation t frequency of genotype *AA*,

$$G_t(\text{A: } AA) = G_t(\text{B: } AA) = D_t.$$

The probability of a $AA \times AA$ mating was:

$$Pr_t(AA \times AA) = G_t(\text{A: } AA)G_t(\text{B: } AA) = D_t^2.$$

Suppose, however, that we had applied a mating rule that specified that the probability that B was of a given genotype depended on the genotype of A. Then the probability of a given mating would involve a conditional probability, for example,

$$Pr_t(AA \times AA) = G_t(\text{A: } AA)G_t(\text{B: } AA \mid \text{A: } AA) \tag{30.1}$$

Mating would no longer be random, but some form of assortative mating, depending on the values assigned for the conditional probability in Eq. (30.1).

Several classes of assortative mating are possible. Under positive assortative mating by genotype, B must be of the same genotype as A. Under positive assortative mating by phenotype, B must be the same phenotype as A; if *A* is dominant to *a*, when A is *AA*, B can be either *AA* or *Aa*, but cannot be *aa*. Negative assortative mating reverses these situations. Under negative assortative mating, B cannot be of the same genotype as A, if assortment is by genotype, or of the same phenotype as A, if it is phenotypic.

Some authors (e.g., Falconer 1981), use the term assortative mating to refer only to positive assortative mating, calling negative assortative mating "disassortative". I have preferred to describe any situation where the choice of B is dependent on the result of choosing A as assortative mating, using positive and negative to indicate the type of assortment.

We will consider only two classes of assortative mating, complete positive assortative mating by genotype and complete negative assortative mating by phenotype, in much detail. Conditional probabilities for the genotype of B given the genotype of A under the various classes of assortative mating are shown in Table 30.1, contrasted with the random mating probabilities.

Under complete positive assortative mating by genotype,

$$G_t(\text{B: } AA \mid \text{A: } AA) = 1,$$

Table 30.1. Conditional probabilities for the genotype of mate B

Genotype of mate		Random mating frequency	Probability under assortative mating			
			Positive by		Negative by	
A	B		Genotype	Phenotype	Genotype	Phenotype
AA	*AA*	D_t	1	$D_t/(D_t+2H_t)$	0	0
	Aa	$2H_t$	0	$2H_t/(D_t+2H_t)$	$2H_t/(D_t+2H_t)$	0
	aa	R_t	0	0	$R_t/(2H_t+R_t)$	1
Aa	*AA*	D_t	0	$D_t/(D_t+2H_t)$	$D_t/(D_t+R_t)$	0
	Aa	$2H_t$	1	$2H_t/(D_t+2H_t)$	0	0
	aa	R_t	0	0	$R_t/(D_t+R_t)$	1
aa	*AA*	D_t	0	0	$D_t/D_t+2H_t)$	$D_t/(D_t+2H_t)$
	Aa	$2H_t$	0	0	$2H_t/(D_t+2H_t)$	$2H_t/(D_t+2H_t)$
	aa	R_t	1	1	0	0

Tabelle 30.2. Positive assortative mating by genotype

Mating	Frequency	*AA*	*Aa*	*aa*
$AA\times AA$	D_t	D_t		
$Aa\times Aa$	$2H_t$	$H_t/2$	H_t	$H_t/2$
$AA\times AA$	H_t			R_t
		$D_t+H_t/2$	H_t	$H_t/2+R_t$

Table 30.3. Results of positive assortative mating by genotype

Generation	*AA*	*Aa*	*aa*
t	D_t	$2H_t$	R_t
t+1	$D_t+H_t/2$	H_t	$R_t+H_t/2$
t+2	$D_t+3H_t/4$	$H_t/2$	$R_t+3H_t/4$
t+3	$D_t+7H_t/8$	$H_t/4$	$R_t+7H_t/8$
t+n	$D_t+[1-(1/2)^n]H_t$	$(1/2)^{n-1}H_t$	$R_t+[1-(1/2)^n]H_t$
Limit	$D_t+H_t=p_t$	0	$R_t+H_t=q_t$

and, assuming random choice of A, the probability of a *AA* x *AA* mating in generation t would be

$$\Pr_t(AA\times AA)=G_t(\text{A: } AA)G_t(\text{B: } AA|\text{A: } AA)$$
$$=D_t(1)=D_t.$$

$Aa\times Aa$, and $aa\times aa$, are the only other two types of matings possible under positive assortative mating by genotype. The probability of a $AA\times Aa$ mating, for example, would be zero, because the conditional probability

$$G_t(\text{B: } Aa|\text{A: } AA)=0.$$

The frequencies of the possible matings, and of the offspring which they produce, are shown in Table 30.2.

Because $Aa \times Aa$ matings produce 0.25 AA, 0.5 Aa and 0.25 aa offspring, while two like homozygotes mated together produce only homozygotes of the same type, the proportion of homozygotes increases rapidly under positive assortative mating by genotype. Table 30.3 shows the results of such matings in generation t, the progression of genotypic frequencies over several generations, and the genotypic frequencies in the limit. Allele frequencies do not change. The frequency of A alleles in generation t+n is

$$
\begin{aligned}
p_{t+n} &= D_t + [1-(1/2)^n]H_t + (1/2)(1/2)^{n-1}H_t \\
&= D_t + [1-(1/2)^n + (1/2)^n]H_t = D_t + H_t = p_t.
\end{aligned}
$$

As n becomes very large, $(1/2)^n$ and $(1/2)^{n-1}$ approach 0, so that, in the limit, complete positive assortative mating by genotype produces a population consisting of p_t AA homozygotes and q_t aa homozygotes, all heterozygotes being eliminated. This is the same result that we obtained from inbreeding. Inbreeding involves mating between related individuals, and, because related individuals are likely to have similar genotypes, can be regarded as a species of positive assortative mating. With assortative mating, however, it is possible to have a population consisting of AA and aa homozygotes with no heterozygotes.

Positive assortative mating by phenotype will also increase the frequency of homozygotes, and, if complete, will lead eventually to a population of p_t AA, q_t aa, no Aa. Partial positive assortative mating, either by phenotype or genotype, will increase homozygote frequency. If there is only a small excess of matings between like individuals over the proportion expected under random mating, the increase in homozygotes will be small (Crow and Kimura 1970). But a large excess of such matings could potentially result in a significant increase in the frequency, say, of a recessive genetic disease in man, if the disease locus were associated with some factor tending to promote assortative mating.

Next let us consider complete negative assortative mating by phenotype. Individuals of the same phenotype cannot mate together; assuming A to be the dominant allele, all matings must be $AA \times aa$ or $Aa \times aa$. Only Aa and aa, in the ratio $D_t + H_t$: H_t, will be produced in generation t+1.

All matings in generation t+1 must be $Aa \times aa$, and will produce Aa and aa offspring in equal numbers. The genotypic ratio in generation t+2 and in all subsequent generations will be 0: 1: 1; equilibrium has been established.

At equilibrium, the frequency of A alleles will be 0.25, that of a, 0.75, regardless of the initial allele frequencies in generation t. The initial allele frequencies and genotypic ratios are entirely irrelevant to the final composition of the population. If dominant phenotypes outnumber recessives in the initial population, either some dominants will be unable to find mates, or some recessives will each mate with several dominant individuals. If recessives outnumber dominants originally, the opposite will be true. In either case, there will be differential fertility among the genotypes, that is to say, selection.

There will also be selection in generation t+1. If there were any AA individuals in generation t, more Aa than aa will be produced in t+1; again, some of these excess *Aa* individuals will be unable to mate unless recessive homozygotes mate with more than one dominant each. Thus, there will be differential fertility and selection. Selection is inherent in complete negative assortative mating by phenotype. The result of this selection will be to change initial gene frequencies to the equilibrium frequencies, $p_e = 0.25$, $q_e = 0.75$.

As Crow and Kimura (1970) point out, the XY sex determining mechanism can be regarded as an example of complete negative assortative mating. The "alleles" are the X and Y chromosomes; mating can occur only between XX and XY; and at equilibrium XX and XY occur in equal numbers, with three fourths of the chromosomes X, one fourth Y. In this case, the YY homozygote does not exist even initially; it would not be produced later.

Negative assortative mating by genotype encompasses a number of different possible situations. Perhaps the most common in nature is self-sterility in plants; one or more genotypes are incapable of successfully mating with another individual of the same genotype (but can mate with any other genotype). Other situations are, however, at least theoretically possible; for example, a given genotype can mate successfully with only one other genotype. Various authors (e.g., Falk and Li 1969, Finney 1952) have explored specific negative assortative mating situations in detail. We will not do so in this text. We will merely note that negative assortative mating can lead to changes in both allelic and genotypic frequencies, and that it generally leads to equilibria at intermediate allelic frequencies.

Lecture 30 Exercises

1. Calculate the genotypic distribution for the first three generations of positive assortative mating by genotype, starting from an initial population in Hardy-Weinberg equilibrium, with $p_t = 0.6$. What is the genotypic distribution in the limit?

2. Starting with a population in Hardy-Weinberg equilibrium, calculate the genotypic and allelic frequencies in the first three generations under negative assortative mating by phenotype. Assume that $p_t = 0.25$, $q_t = 0.75$, and that *A* is dominant to *a*.

Part V Quantitative Inheritance

Lecture 31 Metric Traits

Thus far, we have paid little heed to phenotypes. Our attention has centered on alleles and genotypes, and on the factors affecting their frequencies. Nonetheless, certain assumptions about the nature of phenotypes have been implicit in our discussions.

We have tacitly assumed that genotypes determined phenotypes, so that individuals that differed in phenotype did so because they differed in genotype. (But different genotypes did not necessarily differ in phenotype; for example, *AA* and *Aa* had the same phenotype if *A* was dominant to *a*.)

Also, we have assumed that phenotypic differences were qualitative, that an individual of one phenotype will have some quality or qualities not present in individuals of another phenotype. For example, taster humans have the quality of being able to taste PTC; nontasters do not have this quality. The phenotype of an individual is a qualitative group into which it can be classified, differentiating it from individuals in other qualitative groups. The identification of an individual's phenotype amounts to the classification of that individual. The existence of an occasional individual that is difficult to classify does not vitiate the general rule.

Many important traits of plants and animals, for example, those with which Mendel worked, conform to this picture of qualitative inheritance. The examples we have cited, PTC taste sensitivity, PKU, coat color in mice, etc., have generally been qualitative. But there are also a large number of traits, some of them of surpassing importance to the plant or animal breeder, in which differences between individuals are quantitative, rather than qualitative.

Quantitative traits are measured, rather than being classified; we will refer to them as metric traits, to avoid any confusion between the words quantitative and qualitative. Individuals are characterized for a metric trait by the amount of some quality which they possess, rather than its presence or absence. Examples of traits of importance to plant and animal breeders which are metric rather than classifiable are crop yields, body size in animals, egg production in domestic fowl, and milk production in dairy cattle.

Stature in humans is a more or less typical metric trait. All humans have height, and we differentiate among them by the amount of height that they have. We may say that a short person "lacks height", but this is merely a figure of speech. We do not really mean that such a person lacks the quality of height, only that he has less of this quality than other persons.

We can use human stature to examine the nature of metric traits in more detail. Suppose that we were to measure the heights of a large number of human males of a fairly uniform age, say, the male undergraduate students at Purdue University. We will exclude the young ladies, not from sexual chauvinism, but be-

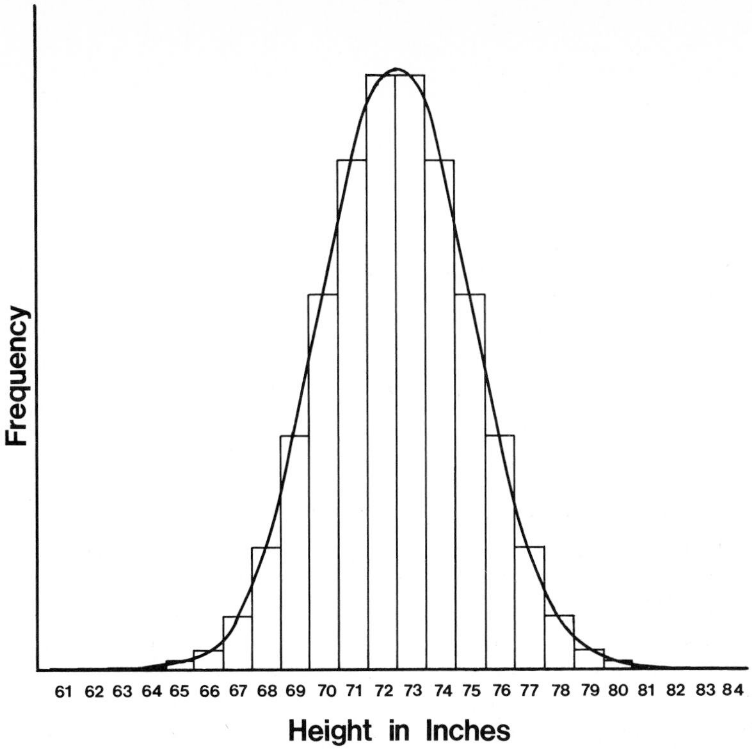

Fig. 31.1. Expected distribution of male stature. The bar graph shows the approximate expected frequencies of one inch height classes. A Gaussian normal distribution is fitted to the bar graph

cause their height distribution differs from that of males. The tallest member of our sample might be a 7-foot basketball player; the shortest might measure in the vicinity of 5 feet. Individuals near these extreme values of the range of heights would be rare, however; those with heights near the middle of the range would be much more common. If we were to measure to the nearest whole inch, and draw a bar graph of the number of individuals in the sample for each one inch height class, the resulting graph would probably look much like that in Fig. 31.1.

This graph approximates the bell-shaped "normal" probability curve described by the nineteenth century mathematician, Gauss, shown superimposed on the bar graph in Fig. 31.1. The Gaussian curve is, of course, continuous, while our graph of statures is discontinuous. The discontinuity is artificial, imposed by classifying the measurements. Classified to the nearest half inch, the student height data would approximate Gauss's distribution more closely. Given an infinite population to measure, and the ability to measure to an infinitesimal fraction, the normal distribution would almost perfectly describe the distribution of heights.

For a qualitative trait, classes occur naturally. It would have been wrong, as well as foolish, if, for example, Hulbert and Doolittle (1971) had classified heterozygous females showing only slight effects of the hair growth mutation together with normal homozygotes. These two phenotypic groups are in different natural phenotypic classes, "some effect (though slight)", and "no effect". (It is, of course, entirely possible, indeed probable, that some heterozygotes showed no discernible effect, and were therefore genetically misclassified.)

With a continuously distributed metric trait, however, there are no natural classes. We may think of one extreme of the height distribution as consisting of tall persons, the other of short. But if we were to ask two people to classify the distribution, they might very well disagree as to where to draw the boundaries between "tall" and "short". They might also disagree as to the appropriate number of classes into which to divide the data. One might, perhaps, argue for three classes, tall, medium, and short, while the other might want to add very tall and very short classes at the extremes. Neither would be right, and neither wrong. Because there are no natural classes, any division is arbitrary. No reasonable division is more or less correct than any other.

This is not to say that arbitrarily dividing a continuous distribution into classes cannot be useful. In the social sciences, classification has been very fruitful, e.g., the concept of socio-economic classes, even though there may be disagreement about the appropriate number of classes and the positions of class limits. But the classification must be appropriate both to the trait classified and to the purposes classification is designed to serve.

For genetic purposes, an appropriate classification must divide phenotypes into classes that at least roughly correspond to different genotypes. We have seen that this is true for the natural classes which characterize qualitative traits. But for the majority of metric traits, it is not possible to divide the phenotypic distribution into classes that correspond to genotypic groups.

For example, attempts to detect segregation ratios in egg production in domestic fowl (Goodale and McMullen 1919) led to the identification of component traits (maturity, rate of lay, persistency, etc.; see Goodale 1918, Goodale and Sanborn 1922). But when Hays (1924) attempted to analyze these component traits by classification methods, he failed because his arbitrary phenotypic classes did not correspond to genotypic groups.

The importance of understanding the inheritance of metric traits is indicated by the importance of the metric traits we have listed above. The dairy breeder must know the genetics of milk production to most effectively improve this trait. Metric traits are affected by genetics, as familial resemblance clearly indicates. Offspring tend to vary from the population average in the same direction as did their parents. But the resemblance is far from exact, and requires a genetic explanation which is more complicated than the simple, traditional, "like begets like" concept.

A theory of metric inheritance based on Mendelian inheritance, the Hardy-Weinberg law, and statistical concepts, has been developed during the first half of the twentieth century. The basic concept of gene action under this theory was set forth by East (1910), when he proposed three duplicate genes to control numbers of rows of kernels per ear in maize. In the presence of any one of these genes, four extra rows would occur. Any two of them would increase the number of rows by eight, while all three together would cause twelve extra rows.

With this type of gene action in mind, the theory proposes that a metric trait is inherited by the action of a large number of gene loci. At each locus, two alleles segregate, one of which tends to increase, the other to decrease, the quantitative expression of the trait. (We will refer to these as plus and minus alleles, respectively.) The loci are duplicate in the sense that each can have either a positive or a negative effect on the trait, depending on the alleles present at that locus. The effects of different loci need not, however, be quantitatively duplicate; one locus may have a greater, another a lesser, effect upon the trait.

All, however, are regarded as having relatively small effects in comparison to the total range of values for the trait seen in the population at large. Large differences between individuals can arise because these individuals differ in genotype at each of a large number of loci.

Typically, the genes involved in qualitative inheritance have large effects, sufficiently large to cause qualitative differences between genotypes. These are major genes; the genes involved in metric inheritance are minor genes, with small effects. The difference between major and minor genes may perhaps be best illustrated by variation in body size. A distribution continuous over some range of body sizes, which can be attributed to metric inheritance, can be observed in nearly every animal species. That is, differences within that range are inherited by a polygenic system of genes which individually have only small effects on body size. But in many species, dwarfing genes are also known. These have a qualitative effect on body size; dwarfs are so much smaller in body size than normal individuals that they fall into a different phenotypic class. Dwarfing genes are major genes; they cause qualitative differences in stature. The genes that cause variation in body size among normal individuals (and among dwarfs as well) are minor genes that also affect stature; but a single locus causes only a small change in the trait.

With the sole exception of the amount of effect that they have, major and minor genes do not differ in any way. Minor genes are of the same physico-chemical nature as major genes. They are transmitted like major genes; they are carried on chromosomes and are subject to linkage; and in all other aspects are the same as major genes.

This doctrine is very important, for it allows us to apply all our knowledge of transmission genetics, biochemical genetics, and particularly population genetics, built up in studies on major genes, to minor genes and metric inheritance. Geneticists, in thinking about a metric trait, assume that the minor genes controlling this trait behave according to the rules we have set forth over the first 30 lectures of this course. This is why a thorough grounding in population genetics is essential to the understanding of metric inheritance.

The small effects and duplicate nature of metric inheritance make it virtually impossible to distinguish the effects of a single locus. Therefore, it is impossible to identify genotypes.

Suppose, for example, that alleles H_1 and h_1 segregate at one of the autosomal loci controlling human stature. H_1, let us imagine, increases stature by 1 mm; h_1 decreases it by a like amount. Let us suppose for the moment that the effect of any genotype is the sum of the effects of the alleles in that genotype. The effects of genotypes H_1H_1, H_1h_1 and h_1h_1 are then $+2$, 0, and -2 mm, respectively.

Next, suppose that we also have a duplicate locus, with alleles H_2 and h_2, having the same effects as H_1 and h_1. The combined effects of the two loci are shown in Table 31.1. Even with only two loci, we can see that genotypes at either locus are beginning to lose phenotypic identity. H_1H_1 can occur in genotypes with +4, +2 or 0 mm effects, according to the second locus genotype with which they are associated. On the other hand, there are two genotypes that can have a +2 mm effect, two an effect of −2 mm, and three a 0 effect.

If we add a third locus (Table 31.2), H_1H_1 can occur in genotypes with effects ranging from −2 to +6 mm; and, except for the two extreme values, at least three genotypes can cause any specific effect. Thus, even with the simplest model of gene action, which we have assumed in the above examples, any given genotypic value can represent several combinations of genes, and a given genotype at one locus can be associated with a range of genetic effects and phenotypic values. If we assume more complex gene action models, such as dominance at one or more loci, or interactions between loci, the association of genotypes with phenotypes becomes even more complex.

Table 31.1. Inheritance of stature; theoretical case

Genotype	Genotypic effect	Including environmental effect
One locus		
H_1H_1	+2 mm	
H_1h_1	0 mm	
h_1h_1	−2 mm	
Two loci		
$H_1H_1H_2H_2$	+4 mm	+2 to +6 mm
$H_1H_1H_2h_2$	+2 mm	0 to +4 mm
$H_1h_1H_2H_2$	+2 mm	0 to +4 mm
$H_1H_1h_2h_2$	0 mm	−2 to +2 mm
$H_1h_1H_2h_2$	0 mm	−2 to +2 mm
$h_1h_1H_2H_2$	0 mm	−2 to +2 mm
$H_1h_1h_2h_2$	−2 mm	−4 to 0 mm
$h_1h_1H_2h_2$	−2 mm	−4 to 0 mm
$h_1h_1h_2h_2$	−4 mm	−6 to −2 mm

Table 31.2. Genotypes and effects for three loci

Genotypes	Effects
$H_1H_1H_2H_2H_3H_3$	+6 mm
$H_1H_1H_2H_2H_3h_3$	+4 mm
$H_1H_1H_2H_2h_3h_3$	+2 mm
$H_1H_1H_2h_2H_3H_3$	+4 mm
$H_1H_1H_2h_2H_3h_3$	+2 mm
$H_1H_1H_2h_2h_3h_3$	0 mm
$H_1H_1h_2h_2H_3H_3$	+2 mm
$H_1H_1h_2h_2H_3h_3$	0 mm
$H_1H_1h_2h_2h_3h_3$	−2 mm
$H_1h_1H_2H_2H_3H_3$	+4 mm
$H_1h_1H_2H_2H_3h_3$	+2 mm
$H_1h_1H_2H_2h_3h_3$	0 mm
$H_1h_1H_2h_2H_3H_3$	+2 mm
$H_1h_1H_2h_2H_3h_3$	0 mm
$H_1h_1H_2h_2h_3h_3$	−2 mm
$H_1h_1h_2h_2H_3H_3$	0 mm
$H_1h_1h_2h_2H_3h_3$	−2 mm
$H_1h_1h_2h_2h_3h_3$	−4 mm
$h_1h_1H_2H_2H_3H_3$	+2 mm
$h_1h_1H_2H_2H_3h_3$	0 mm
$h_1h_1H_2H_2h_3h_3$	−2 mm
$h_1h_1H_2h_2H_3H_3$	0 mm
$h_1h_1H_2h_2H_3h_3$	−2 mm
$h_1h_1H_2h_2h_3h_3$	−4 mm
$h_1h_1h_2h_2H_3H_3$	−2 mm
$h_1h_1h_2h_2H_3h_3$	−4 mm
$h_1h_1h_2h_2h_3h_3$	−6 mm

The effects of the environment can further obscure differences between genotypes. We can see these effects by assuming, in the two locus case, that environment can change the phenotype of any individual by as much as 2 mm in either direction from the genotypic effect. The resulting phenotypic ranges for each genotype are shown in Table 31.1. The effect of the environment is to increase the range of phenotypic values which any given genotype can produce, and thus further to prevent the phenotypic identification of the genotype from its phenotype.

Lecture 31 Exercises

1. Suppose that in the three locus case, H_1 is dominant to h_1 and H_2 to h_2. That is, H_1h_1 has the same effect as H_1H_1, H_2h_2 as H_2H_2. Furthermore, in the presence of H_1H_1, H_3 is dominant to h_3, due to interlocus interaction. (In the presence of H_1h_1 or h_1h_1, H_3 and h_3 are codominant.) How many genotypes will have identical genotypic values under these conditions?

2. In the presence of a ± 2 mm environmental effect, how many genotypes could show a phenotypic value of $+4$ mm under the conditions of Exercise 1?

Lecture 32 Evidence for the Theory of Metric Inheritance

The history of the development of the theory of metric inheritance has recently been reviewed by Hill (1985). Yule (1902, cited by Hill 1985) appears to have been the first to suggest polygenic inheritance of metric traits. Pearson (1904) provided the first mathematical treatment of a polygenic model. Fisher (1918) introduced the concept of partitioning genotypic variance into additive and dominance effects, noting that full sib covariance included both while covariance between parent and offspring included only additive effects. Wright (1920, 1921) independently derived an analysis of metric inheritance, also partitioning additive and dominance genotypic effects, and introducing the concepts of broad sense and narrow sense heritability. Wright also analyzed environmental effects into those common to relatives and those specific to individuals, as suggested by Weinberg (1910).

At the heart of this theory is the concept that major and minor genes do not differ in any way except in the amount of their effect. Through eight and a half decades of study, we have learned a great deal about the transmission, action, and even the physical and chemical nature of major genes. The theory of metric inheritance assumes that minor genes are transmitted, and act, in the same way as major genes. Predictions of the results of genetic manipulations are based on this assumption. Thus, any evidence we can obtain indicating that this assumption is true is evidence supporting the theory.

There is considerable evidence that major and minor genes can both affect the same trait. Body size is one of the most frequently manipulated metric traits among animals. Whether measured in terms of dimensions (e.g., human stature) or in terms of body weight, it is clearly a metric trait, and obviously of great importance in breeding animals for increased production of meat. Some of the earliest experimental studies of selection for a metric trait were those of Goodale, starting in the early 1930's (Wilson et al. 1971) and of MacArthur (1949) starting toward the end of that decade. Both studies used adult body weight of mice as the selection criterion.

Yet this classic metric trait is also subject to qualitative variation. As we have already noted, in nearly every species that has been studied intensively, major genes have been found that cause qualitative differences in body size, the dwarfing genes.

Both normal and dwarf individuals vary in body size, under the effects of other genotypic factors and of environment. But the dwarfing gene causes a difference between dwarf and normal homozygotes so large that it is not obscured by other variation. The dwarfing gene's effect is qualitative merely because it transcends variation due to other causes. If other variation were sufficiently large, or the dwarfing gene's effect sufficiently small, this gene would be regarded as another minor gene contributing to a metric trait. Or, to put it the other way around, minor genes for decreased body size are dwarfing genes whose effects are too small to be distinguished qualitatively. Major and minor genes are part of the same continuum of effects.

Garrod's *Inborn Errors of Metabolism* (1909) first suggested that genes might act through the control of metabolism. Clearly, the metabolic differences Garrod investigated were qualitative differences caused by major genes. At the other end of the spectrum, isoalleles are known to cause differences in enzyme structure which do not alter enzyme function. Between these extremes, genes are known that alter enzyme function only slightly, perhaps changing the optimum pH or temperature conditions for the enzyme. Such minor enzyme changes would presumably result in very small changes in metabolic rates under conditions prevailing in the organism. For example, the pH of stomach contents may be optimum for the function of a normal gastric enzyme, but an altered enzyme might have a different optimum pH and therefore might function at a slower rate under normal stomach conditions. This decreased digestive capacity might be reflected in a slower growth rate for the animal producing the altered enzyme, and thus in a lower body weight at a given age. Bell's (1982) suggestion, previously mentioned, that minor gene mutation rates may be higher than those for major genes, may indicate that minor gene mutations involve less profound changes in the genes.

All of this suggests that both major and minor gene changes may occur at a given locus, further evidence that major and minor genes differ only in the amount of effect they exert.

A single minor gene cannot usually be identified because its effects are masked by the variation caused by environment and by other genes. Doolittle (1961), however, identified minor gene effects on a metric trait of genes with major qualitative effects on pelage characteristics in the mouse. Thus, the same gene can have both a major gene effect on one trait and a minor effect on another, certain evidence that amount of effect is the only difference.

Minor genes work together in polygenic systems to produce large overall differences between individuals. Presumably, if only a few minor gene loci affect a trait, the genetic influence on that trait would never be identified; genotypic differences would be so slight as to escape detection. Thus, the polygenic nature of minor gene inheritance is, to some extent, artificial. A single major gene locus, by definition, has an effect large enough to be noticed. But major genes can also work together in polygenic systems. Silvers (1979) cited more than 60 gene loci affecting coat color in the mouse. Each has a specific effect on one or more of the attributes which comprise coat color (pigment color, intensity, distribution, etc.). Overall coat color is the result of the combination of these genes present in a given individual, a polygenic system.

Similarly, a large number of antigenic differences have been reported in the erythrocytes of man, not only the ABO, MNS, and Rh systems, but less well-known systems such as Lewis, Lutheran, etc. Each antigenic system represents a genetic locus, with qualitative, major gene effects. The whole complex of loci that control these various antigens make up a polygenic system controlling the antigenic surface of the human erythrocyte.

Major genes controlling qualitative traits exhibit dominance and epistasis, the latter being the situation where the expression of the genotype at one locus is dependent on the genotype at another. As we shall shortly see, metric traits inherited by minor genes exhibit variations in expression which can be attributed to phenomena analogous to dominance and epistasis.

Thus, there is considerable evidence that minor genes, if they exist at all, are no different than major genes, excepting always in the amount of their effect. Direct evidence of the existence of minor genes is less easy to obtain. The whole circumstances of minor gene inheritance militate against the possibility of obtaining direct evidence. Perhaps the best such evidence is the demonstration by Doolittle (1961), mentioned above, that identifiable major genes can have minor gene effects on a metric trait.

The most compelling evidence for the theory as a whole, however, is simply that it works. It does predict successfully the results of observations on natural populations of living creatures, and those of genetic manipulation of metric traits in domestic populations. Discrepancies between experimental results and theoretical predictions usually can be shown to be due to the failure to take into account or correctly evaluate the effects, in a minor gene setting, of some known aspect of major gene inheritance. Such discrepancies do not require a basic reformulation of the entire theory. Thus we can proceed to discuss metric inheritance in terms of this theory with reasonable confidence.

Lecture 33 Phenotypic Value

The phenotype of an individual for a qualitative trait is the class into which that individual's phenotype falls. But metric traits are measured, rather than classified. The phenotype of the ij-th individual in a population with regard to a given metric trait is the value of the trait measured in individual ij, P_{ij}.

The geneticist regards the phenotypic value of an individual as the summation of the effects of its genotype and its environment. We can write this model in its simplest form as

$$P = G + E;$$

the phenotypic value is the sum of the genotypic and environmental effects. It is convenient to use the statistician's ploy of expressing both genotypic and environmental effects as deviations from the mean of a theoretical population of individuals expressing all possible combinations of genotypes and environments. Thus, a more useful form of the model can be written as

$$P_{ij} = \mu + G_i + E_{ij}, \qquad (33.1)$$

where P_{ij} is the phenotypic value measured on the j-th individual with genotype i; μ is the general mean; G_i is the effect of the i-th genotype; and E_{ij} is the effect of the environment on individual ij.

In our model, we lump together all factors causing the phenotypic value of an individual to differ from the population mean, except the direct effects of the genotype, and label the sum of these effects environmental. E_{ij} is a wastebasket category, representing the difference between the phenotypic value and the sum of μ and the genotypic effect. It may include factors which an ecologist, say, would hesitate to classify as environmental. Macro- and micro-environmental factors currently impinging on individual ij, for example, the current temperature, availability of nutrients, etc., are certainly included in E_{ij}. But so are factors that have affected the individual at any time in its developmental history prior to the measurement of the phenotypic value, if these historical effects have carried over to the present. The geneticist may even, under some circumstances, include genotypic effects in E_{ij}. For example, suppose that mouse ij carries a gene *X* which affects its body size, and that gene *X* was inherited from ij's dam. *X* will have a direct effect on ij's body size, which will be included in G_i. But *X* will also have affected the body size of ij's dam, and thus the quality of the intrauterine environment she afforded her pups. The intrauterine environment may well have effects on body size which carry over to the offspring's mature body size. These maternal effect will be regarded as environmental effects on ij, and included in E_{ij}.

E_{ij} is the sum of the effects of various factors, some of which will be positive and others negative. The sum of these effects may be either positive or negative for a particular individual. Because E_{ij} is the difference between P_{ij} and $\mu+G_i$, however, its expected value is 0 in an infinite population. For every individual whose $E_{ij}=+x$, there will be another with $E_{i*j*}=-x$. Among the individuals in a finite sample, of course, this will not necessarily be true.

$$\bar{P}_{i.} \simeq \mu+G_i$$

would be a valid approximation if $\bar{P}_{i.}$ were based upon a sufficiently large number of individuals, all of genotype i.

The model assumes that there are no genotype x environment interaction effects. If there were, an additional term would have to be added to the model:

$$P_{ij}=\mu+G_i+E_{ij}+(GE)_{ij}.$$

Genotype x environment interaction is often confused with genotype-environment correlation. The latter occurs if genotypic and environmental effects are not independent. It does not alter the form of the linear model, but when we derive the variance of phenotypic values from the model, a correlation will lead to the appearance of a genotype-environment covariance. We will introduce the concept of a genotype-environment correlation here, out of its proper context, in order to make clear the contrast between correlation and interaction.

Briefly, there is an interaction if the difference between the average phenotypic values for two genotypes changes in different environments. There is a correlation if particular genotypes tend to be associated with positive, and other genotypes with negative, environmental effects. Let us develop these two concepts in more detail.

Suppose that we have two genotypes, i and i*, i having the greater genotypic value. If there are no genotype x environment interactions, $\bar{P}_{ij}-\bar{P}_{i*j}$ will be the same in all environments. If such an interaction occurs, $\bar{P}_{ij}-\bar{P}_{i*j}$ will differ from one environment to another.

The breeder must be concerned about such interactions, because the environment in which he produces and selects his stock may be rather different from that on producers' farms. For example, suppose that Breeders A and B are competing for the market for meat rabbit stock. Breeder A surreptitiously purchases B's stock and compares it to his own. He finds that his stock grows considerably faster than B's, but has a slightly lower feed efficiency, under the environment he provides. If there is no genotype x environment interaction effect on growth rate and feed efficiency, breeder A can be confident that rabbit meat producers will find similar differences in the environments that they provide, so that many will prefer his stock.

However, a gentoype x environment interaction effect on feed efficiency might greatly exaggerate breeder B's advantage in this regard, on producers' farms. An interaction effect on growth rate might reduce A's advantage in this trait; it might even reverse the ranking of the two stocks. On producers' farms, B's stock might grow faster.

Major genotype x environment interaction effects appear to be relatively rare among economic species of animals, although more common in plants. In general, while the differences between two plant stocks may be much greater in one environment than in another, changes in ranking of the stocks seldom occur. The stock superior in one environment will usually be superior in the other also.

We usually assume that genotype and environment are uncorrelated, but the validity of this assumption has often been questioned. The classic counterexample is the feeding of grain concentrates to dairy cattle in proportion to the amount of milk which they produce. The amount of milk produced is, of course, determined at least in part by the cow's genotype. Therefore, cows that have a better genotype for milk production receive more concentrates, a better environment. Genotype and environment are positively correlated.

The fallacy in this argument lies in a confusion between environment and environmental effect. We assume in our model independence of genotypic and environmental effects, not necessarily between genotype and environment, per se. The cow producing more milk does receive a better environment, more concentrates; but she also expends extra nutrients to produce extra milk. The purpose, and by and large the effect, of the extra concentrates given the high-producing cow is to replace the extra expenditure. It can be argued that the environment of the high-producing cow is thus balanced to a level similar to that of the low producer; failing to replace the extra expenditure of nutrients by the high producer would, in fact, create a negative correlation of genotypic and environmental effects.

The G_i are measured as deviations from the mean, μ, and therefore they also, in an infinite population, sum to 0. The mean of the G_i in a large finite sample can therefore be taken as approximately 0. As a result, if there are no correlations or interactions between phenotypic and genotypic effects, the average of all phenotypic values in a large population can be taken as an estimate of μ. But μ can be interpreted as the value of the average individual from the population, who has the average genotype.

We have already pointed out that if we take a large group of individuals, all of genotype i, environmental effects will cancel out of the average of their phenotypic values, if genotype and environment are uncorrelated. This average will then estimate $\mu + G_i$. The difference between the average of individuals of genotype i and that for all individuals in the population is an estimate of G_i.

The flaw in this procedure for estimating genotypic effects is that genotypes for metric traits are not distinguishable. Because we cannot identify individuals with the same genotype, we cannot measure phenotypic values for a large number of individuals of genotype i, and use their mean phenotypic value to estimate G_i, except in special circumstances such as completely inbred lines.

Lecture 34 The Single Locus Model

We cannot use the simple quantitative model,

$$P_{ij} = \mu + G_i + E_{ij},$$

to estimate genetic effects on a metric trait. We can measure P_{ij}, the phenotypic value of individual ij, but we cannot identify the genotype of that individual. Many loci with quasi duplicate effects act upon the trait, so that any of a number of different genotypes may produce the same G_i value. The genotype is further obscured by environmental variation, so that the same P_{ij} may result from any of a number of different G_i values.

Nonetheless, G_i is the result of the effects of genotypes at individual loci. Although we cannot measure these, we can at least explore in theory the effects of the genotypes at a particular locus. Suppose, then, that we have two alleles, A_1 and A_2, segregating at an autosomal locus in a diploid. These genes affect some metric trait; the genotypes A_1A_1, A_1A_2, and A_2A_2 have genotypic values Y_{11}, Y_{12}, and Y_{22}, respectively, on whatever scale of values we use to measure the trait: body weight in pounds or kilograms, crop yield in kilograms per hectare, etc. Because we have specified that the A locus affects the trait, Y_{11}, Y_{12} and Y_{22} cannot all be equal to one another. Let us further suppose that $Y_{11} > Y_{22}$. This does not affect the generality of our argument; we are merely agreeing to call the plus allele A_1, and the minus allele A_2.

We can now define a quantity, **a**, as half the difference between Y_{11} and Y_{22}:

$$\mathbf{a} = (Y_{11} - Y_{22})/2. \qquad (34.1)$$

The average of Y_{11} and Y_{22} is, of course,

$$\tilde{y} = (Y_{11} + Y_{22})/2.$$

This average is equidistant from the two homozygotes, being **a** units less than Y_{11} and **a** units greater than Y_{22}:

$$Y_{11} - \tilde{y} = Y_{11} - (Y_{11} + Y_{22})/2 = (2Y_{11} - Y_{11} - Y_{22})/2$$
$$= (Y_{11} - Y_{22})/2 = \mathbf{a};$$
$$Y_{22} - \tilde{y} = (2Y_{22} - Y_{11} - Y_{22})/2 = -\mathbf{a}.$$

The Y values are measured on an absolute scale whose zero point will usually lie far outside the range Y_{22} to Y_{11}. We can transpose Y_{11} and Y_{22} onto a relative scale of values, which uses the same units as the absolute scale, but in which the zero point is at $\tilde{y}$. (This is accomplished by subtracting $\tilde{y}$ from Y_{11} and Y_{22}.) The effects of genotypes A_1A_1 and A_2A_2 are then, respectively, $+\mathbf{a}$ and $-\mathbf{a}$. Figure

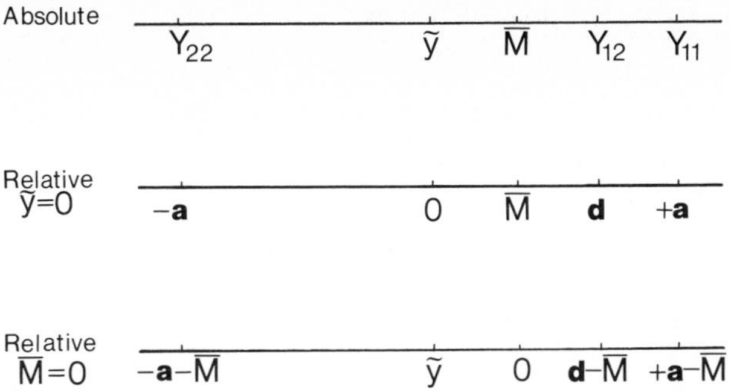

Fig. 34.1. The three scales of measurement. On the absolute scale, $\tilde{y}=(Y_{11}+Y_{22})/2$, $\bar{M}=p^2Y_{11}+2pqY_{12}+q^2Y_{22}$. On the $\tilde{y}$ scale, $\bar{M}=(q-p)\mathbf{a}-2pq\mathbf{d}$. On the $\bar{M}$ scale, $\tilde{y}=(p-q)\mathbf{a}+2pq\mathbf{d}$

34.1 may be helpful in understanding the relationships between these two scales of measurement.

We can also define a quantity **d** as the difference between Y_{12} and the average of Y_{11} and Y_{22}:

$$\mathbf{d}=Y_{12}-\tilde{y}, \tag{34.2}$$

on the absolute scale; on the relative scale with $\tilde{y}=0$, the value of the A_1A_2 genotype will be **d**.

d can, of course, be either positive or negative, and either greater or less than **a** in absolute value. The size and sign of **d** depend on the dominance relationship between A_1 and A_2.

If Y_{12} lies exactly halfway between Y_{11} and Y_{22}, $Y_{12}=\tilde{y}$, and **d** is 0. Neither A_1 nor A_2 is dominant to the other; there is no dominance. If Y_{12} lies between Y_{11} and $\tilde{y}$, $0<\mathbf{d}<\mathbf{a}$, and A_1 is partially dominant to A_2. If Y_{12} coincides with Y_{11}, $\mathbf{d}=\mathbf{a}$; A_1 is completely dominant to A_2. If $Y_{12}>Y_{11}$, $\mathbf{d}>\mathbf{a}>0$, and we have overdominance of A_1, positive overdominance because A_1 is the plus allele. Similarly, if $\tilde{y}>Y_{12}>Y_{22}$, $0>\mathbf{d}>-\mathbf{a}$; if $Y_{12}=Y_{22}$, $\mathbf{d}=-\mathbf{a}$; and if $Y_{22}>Y_{12}$, $0>-\mathbf{a}>\mathbf{d}$. These situations correspond to partial dominance, complete dominance and overdominance, respectively, of the A_2 allele. Because this is the minus allele, overdominance of A_2 is negative overdominance. Table 34.1 summarizes the possible dominance situations; the scale at the bottom of the table shows the positions of Y_{12} for each case.

Because **d** is dependent on the dominance relationship between A_1 and A_2, and is 0 when there is no dominance, it is often referred to as the dominance effect at the A locus.

If we assume a Hardy-Weinberg distribution of genotypes, we can define the population average of the effects of the locus. This average, $\bar{M}$, will be a weighted average of genotypic values, each value being weighted by the frequency of that genotype. Thus,

$$\begin{aligned}\bar{M}&=[p^2\mathbf{a}+2pq\mathbf{d}+q^2(-\mathbf{a})]/(p^2+2pq+q^2)\\&=(p^2-q^2)\mathbf{a}+2pq\mathbf{d}=(p-q)\mathbf{a}+2pq\mathbf{d}\end{aligned} \tag{34.3}$$

Table 34.1. Value of d relative to a for varying levels of dominance

Level	Dominant allele	d
1. Negative overdominance	A_2	$\mathbf{d}<-\mathbf{a}$
2. Complete dominance	A_2	$\mathbf{d}=-\mathbf{a}$
3. Partial dominance	A_2	$0>\mathbf{d}>\mathbf{a}$
4. No dominance	Neither	$\mathbf{d}=0$
5. Partial dominance	A_1	$0<\mathbf{d}<\mathbf{a}$
6. Complete dominance	A_1	$\mathbf{d}=\mathbf{a}$
7. Positive overdominance	A_1	$\mathbf{d}>\mathbf{a}$

	A_2A_2				A_1A_1	
	$-\mathbf{a}$		0		$\mathbf{a}$	
1	2	3	4	5	6	7

We have already introduced two scales on which genotypic values can be measured. The first was the absolute scale, on which the genotypic values were Y_{11}, Y_{12} and Y_{22}. The second was a relative scale, with $\tilde{y}$ as its 0 point; on this scale, the genotypic values were **a**, **d** and $-$**a**.

We will now introduce yet a third scale, with $\bar{M}$ as its zero point. To transform to this scale, we subtract $\bar{M}$ from the genotypic values on the $\tilde{y}=0$ scale. The transforming functions are

$$\begin{aligned} f(Y_{11}) &= \mathbf{a}-\bar{M}=\mathbf{a}-(p-q)\mathbf{a}-2pq\mathbf{d} \\ &= (1-p+q)\mathbf{a}-2pq\mathbf{d}=2q\mathbf{a}-2pq\mathbf{d} \\ f(Y_{12}) &= \mathbf{d}-(p-q)\mathbf{a}-2pq\mathbf{d}=\mathbf{d}+(q-p)\mathbf{a}-2pq\mathbf{d}; \\ f(Y_{22}) &= -\mathbf{a}-(p-q)\mathbf{a}-2pq\mathbf{d} \\ &= -(1+p-q)\mathbf{a}-2pq\mathbf{d}=-2p\mathbf{a}-2pq\mathbf{d}. \end{aligned} \tag{34.4}$$

Note that $f(Y_{12})$ is **d** plus the average of $f(Y_{11})$ and $f(Y_{22})$:

$$\begin{aligned} [f(Y_{11})+f(Y_{22})]/2 &= (2q\mathbf{a}-2pq\mathbf{d}-2p\mathbf{a}-2pq\mathbf{d})/2 \\ &= (q-p)\mathbf{a}-2pq\mathbf{d}. \end{aligned}$$

Thus, the value of the heterozygote on the $\bar{M}$ scale is still **d** units from the average of the values of the two homozygotes.

Table 34.2 and Fig. 34.1 depict the three scales to help clarify their interrelationships. The location of the zero point of the third scale, $\bar{M}$, depends on the values of **a** and **d**, and on the allele frequencies p and q. This, in fact, is the principle difference between the $\tilde{y}$ and $\bar{M}$ scales; $\tilde{y}$ is the unweighted mean of the effects of the two homozygous genotypes, while $\bar{M}$ is the mean of all three genotypes, weighted by their Hardy-Weinberg frequencies.

Individuals inherit genes, not genotypes, from their parents. The plant or animal breeder is more interested in the offspring that his stock can produce than in the stock itself. Therefore, he is interested in the effects of genes, perhaps more than in those of genotypes.

Table 34.2. The three scales of measurement

Genotype or function	Absolute value	Relative value $\tilde{y}=0$	Relative value $\bar{M}=0$
A_1A_1	Y_{11}	$\mathbf{a}$	$\mathbf{a}-\bar{M}$
A_1A_2	Y_{12}	$\mathbf{d}$	$\mathbf{d}-\bar{M}$
A_2A_2	Y_{22}	$-\mathbf{a}$	$-\mathbf{a}-\bar{M}$
$\tilde{y}$	$(Y_{11}+Y_{22})/2$	0	$(p-q)\mathbf{a}-2pq\mathbf{d}$
$\bar{M}$	$p^2Y_{11}+2pqY_{12}+q^2Y_{22}$	$(q-p)\mathbf{a}-2pq\mathbf{d}$	0

We can define the effect of allele A_i in terms of the average value of genotypes in which A_i appears. We will measure the average effect on the $\bar{M}$ scale; therefore, genotypic values on this scale are used, and are weighted by their Hardy-Weinberg frequencies, used in calculating $\bar{M}$. In addition, the genotypic values must be weighted by the number of A_i genes in the genotype. Thus, for example, genotype A_1A_1 contains 2 A_1 alleles; its Hardy-Weinberg frequency is p^2; and its value on the $\bar{M}$ scale is $\mathbf{a}-\bar{M}$. Therefore, $2p^2(\mathbf{a}-\bar{M})$ is the term for this genotype in $M(A_1)$, the average effect of A_1. The sum of the weighted terms, divided by the sum of the weights, yields the weighted average for each allele.

$$\begin{aligned}
M(A_1)&=[(2)(p^2)(\mathbf{a}-\bar{M})+(1)(2pq)(\mathbf{d}-\bar{M})+(0)(q^2)(-\mathbf{a}-\bar{M})]/(2p^2+2pq+0)\\
&=[2p^2\mathbf{a}+2pq\mathbf{d}-(2p^2+2pq)\bar{M}]/[2p(p+q)]\\
&=[2p(p\mathbf{a}+q\mathbf{d})/2p]-\bar{M}=p\mathbf{a}+q\mathbf{d}-(p-q)\mathbf{a}-2pq\mathbf{d}\\
&=q\mathbf{a}+q(1-2p)\mathbf{d}=q[\mathbf{a}+(q-p)\mathbf{d}];\\
M(A_2)&=[2pq\mathbf{d}+2q^2(-\mathbf{a})-(2pq+2q^2)\bar{M}]/(2pq+2q^2)\\
&=p\mathbf{d}-q\mathbf{a}-(p-q)\mathbf{a}-2pq\mathbf{d}\\
&=-p\mathbf{a}+(p-2pq)\mathbf{d}=-p[\mathbf{a}-(1-2q)\mathbf{d}]\\
&=-p[\mathbf{a}+(q-p)\mathbf{d}]. \qquad (34.5)
\end{aligned}$$

$M(A_1)$ and $M(A_2)$ have the term $\mathbf{a}+(q-p)\mathbf{d}$ in common. This term is conventionally symbolized by α, and the average values of the alleles written as

$$M(A_1)=q\alpha,$$

$$M(A_2)=-p\alpha.$$

It should be emphasized that the above definition of α is based on the assumption of random mating and the Hardy-Weinberg ratio.

The effect of gene A_i can also be defined in terms of the change in genotypic value caused by substituting A_i for its allele, A_{i*}. For example, if we substitute A_1 for A_2 in genotype A_2A_2, the change in genotypic value is the value of the new genotype minus that of the old,

$$\mathbf{d}-\bar{M}-(-\mathbf{a}-\bar{M})=\mathbf{d}+\mathbf{a}.$$

Table 34.3. Changes in genotypic value due to gene substitutions

Allele substituted	Allele replaced	Old genotype	New genotype	Old value	New value	Change (new-old)
A_1	A_1	A_1A_1	A_1A_1	$\mathbf{a}-\bar{\mathrm{M}}$	$\mathbf{a}-\bar{\mathrm{M}}$	0
		A_1A_2	A_1A_2	$\mathbf{d}-\mathrm{M}$	$\mathbf{d}-\mathrm{M}$	0
	A_2	A_1A_2	A_1A_1	$\mathbf{d}-\bar{\mathrm{M}}$	$\mathbf{a}-\bar{\mathrm{M}}$	$\mathbf{a}-\mathbf{d}$
		A_2A_2	A_1A_2	$-\mathbf{a}-\mathrm{M}$	$\mathbf{d}-\mathrm{M}$	$\mathbf{d}+\mathbf{a}$
A_2	A_1	A_1A_1	A_1A_2	$\mathbf{a}-\bar{\mathrm{M}}$	$\mathbf{d}-\bar{\mathrm{M}}$	$\mathbf{d}-\mathbf{a}$
		A_1A_2	A_2A_2	$\mathbf{d}-\mathrm{M}$	$-\mathbf{a}-\mathrm{M}$	$-\mathbf{a}-\mathbf{d}$
	A_2	A_1A_2	A_1A_2	$\mathbf{d}-\bar{\mathrm{M}}$	$\mathbf{d}-\bar{\mathrm{M}}$	0
		A_2A_2	A_2A_2	$-\mathbf{a}-\mathrm{M}$	$-\mathbf{a}-\mathrm{M}$	0

Table 34.3 summarizes the changes in genotypic value caused by gene substitutions. Substituting A_1 for A_1, or A_2 for A_2, causes no change in value, of course. In each change, the $\bar{\mathrm{M}}$'s cancel; thus, substitution values are the same on the $\bar{\mathrm{M}}$ and the $\tilde{y}$ scales.

The probability of substituting A_1 for A_2 in genotype A_2A_2, rather than in A_1A_2, is the relative frequency of A_2 alleles in the two genotypes. The frequency in homozygotes is

$$2q^2/(2pq+2q^2)=q.$$

The frequency in heterozygotes is p, and the change in genotypic value from substituting A_1 for A_2 is $\mathbf{a}-\mathbf{d}$. Therefore, the substitution values of the alleles are

$$\begin{aligned} S(A_1)&=p(\mathbf{a}-\mathbf{d})+q(\mathbf{d}+\mathbf{a})\\ &=\mathbf{a}+(q-p)\mathbf{d}=\alpha;\\ S(A_2)&=p(\mathbf{d}-\mathbf{a})+q(-\mathbf{a}-\mathbf{d})\\ &=-p\mathbf{a}+(q-p)\mathbf{d}=-\alpha. \end{aligned} \tag{34.6}$$

Substitution thus defines genic values in terms of the same parameter as average value. There are q A_2 genes in the population; therefore, we can substitute A_1 for A_2 with frequency q. The average effect of the substitution of A_1 for A_2 is

$$qS(A_1)=q\alpha=M(A_1);$$

the average effect of the reverse substitutions,

$$pS(A_2)=-p\alpha=M(A_2).$$

The *hs* gene in mice causes white head spots (Wildman 1984, Wildman and Doolittle 1986). We can use a mating scheme which will allow us to produce individuals whose head-spot locus genotype, *HsHs*, *Hshs*, or *hshs*, can be identified beyond any doubt. Suppose that we weigh a large number of mice of each of the three genotypes and find average weights of 20.855, 20.852, and 20.847 g, respectively. These weights represent estimates of the genotypic values for the three genotypes, environmental variation tending to cancel over a large number of individuals. We want to estimate the genetic effects of the head-spot locus on body weight.

$$\tilde{y}=(Y_{11}+Y_{22})/2=(20.855+20.847)/2=20.851;$$

and we can estimate

$$\mathbf{a}=(Y_{11}-Y_{22})/2=(20.855-20.847)/2=0.004,$$

and

$$\mathbf{d}=Y_{12}-\tilde{y}=20.852-20.851=0.001.$$

In a population in which we have $p=0.6$, where p is the frequency of *Hs*, $\bar{M}$ would be

$$\bar{M}=(p-q)\mathbf{a}+2pq\mathbf{d}=(0.6-0.4)(0.004)+2(0.24)(0.001)$$
$$=0.0008+0.00048=0.00128.$$

The effects of the alleles are measured in terms of α, which is $\mathbf{a}+(q-p)\mathbf{d}$, so that

$$\alpha=0.004-(0.2)(0.001)=0.0038;$$

then

$$M(Hs)=q\alpha=0.4(0.0038)=0.00152;$$
$$M(hs)=-p\alpha=-(0.6)(0.0038)=-0.00228.$$

[Doolittle (1982), found no significant correlation between headspot size and body weight; the effect of the presence or absence of head spots on body weights has not been investigated. It seems unlikely, however, that there is any significant effect; our example is purely fictitious.]

Lecture 34 Exercises

1. At a given locus, we have the following situation:

Genotype	Frequency	Value
A_1A_1	0.49	1
A_1A_2	0.42	−2
A_2A_2	0.09	−3

Evaluate p and q, **a** and **d**, and $\bar{M}$.

2. Evaluate the effects of alleles A_1 and A_2.

Lecture 35 Breeding Values

From the point of view of the plant or animal breeder, an individual's worth lies more in its ability to transmit performance qualities to its offspring than in its own performance. This principle is embodied in the concept of the breeding value of the individual, defined as twice the average value of that individual's potential offspring, taken as a deviation from the population mean. The average is doubled to adjust for the fact that only half of the offspring values are derived from the genes of the parent being evaluated.

The theoretical breeding value for each parental genotype is twice the average value of the offspring which an individual of that genotype would produce if mated to A_1A_1, A_1A_2, and A_2A_2 mates in random mating proportions.

Consider the breeding value of the genotype A_1A_1. If A_1A_1 is mated to A_1A_1, all offspring will be A_1A_1. The value of such offspring on the $\bar{\mathrm{M}}$ scale will be $\mathbf{a}-\bar{\mathrm{M}}$. If A_1A_1 is mated to A_2A_2, all offspring will be A_1A_2, and their value will be $\mathbf{d}-\bar{\mathrm{M}}$. If the mate is A_1A_2, half the offspring will be A_1 homozygotes, half heterozygotes, and their value will be

$$(1/2)(\mathbf{a}-\bar{\mathrm{M}})+(1/2)(\mathbf{d}-\bar{\mathrm{M}})=(1/2)(\mathbf{a}+\mathbf{d})-\bar{\mathrm{M}}.$$

Offspring values are summarized in Table 35.1.

The breeding value of A_1A_1, which we shall symbolize as $\mathbf{A}_{11}$, is

$$\begin{aligned}\mathbf{A}_{11}&=2\{\mathrm{p}^2(\mathbf{a}-\bar{\mathrm{M}})+2\mathrm{pq}[(1/2)(\mathbf{a}+\mathbf{d})-\bar{\mathrm{M}}]+\mathrm{q}^2(\mathbf{d}-\bar{\mathrm{M}})\}\\&=2(\mathrm{p}^2\mathbf{a}+\mathrm{pq}\mathbf{a}+\mathrm{pq}\mathbf{d}+\mathrm{q}^2\mathbf{d}-\bar{\mathrm{M}})\\&=2[\mathrm{p}^2\mathbf{a}+\mathrm{pq}\mathbf{a}+\mathrm{pq}\mathbf{d}+\mathrm{q}^2\mathbf{d}-(\mathrm{p}-\mathrm{q})\mathbf{a}-2\mathrm{pq}\mathbf{d}].\end{aligned}$$

The coefficient of **a** in the above expression is

$$\mathrm{p}^2+\mathrm{pq}-\mathrm{p}+\mathrm{q}=\mathrm{p}(\mathrm{p}+\mathrm{q})-\mathrm{p}+\mathrm{q}=\mathrm{q};$$

that of **d** is

$$\mathrm{pq}+\mathrm{q}^2-2\mathrm{pq}=\mathrm{q}^2-\mathrm{pq}=\mathrm{q}(\mathrm{q}-\mathrm{p}).$$

Table 35.1. Value of potential offspring

Genotype of mate	Frequency	Genotype of given parent		
		A_1A_1	A_1A_2	A_2A_2
A_1A_1	p^2	$\mathbf{a}-\bar{\mathrm{M}}$	$(1/2)(\mathbf{a}+\mathbf{d})-\bar{\mathrm{M}}$	$\mathbf{d}-\bar{\mathrm{M}}$
A_1A_2	$2\mathrm{pq}$	$(1/2)(\mathbf{a}+\mathbf{d})-\bar{\mathrm{M}}$	$(1/2)(\mathbf{d})-\bar{\mathrm{M}}$	$(1/2)(\mathbf{d}-\mathbf{a})-\bar{\mathrm{M}}$
A_2A_2	q^2	$\mathbf{d}-\bar{\mathrm{M}}$	$(1/2)(\mathbf{d}-\mathbf{a})-\bar{\mathrm{M}}$	$-\mathbf{a}-\bar{\mathrm{M}}$

Therefore,

$$\mathbf{A}_{11}=2[\mathrm{q}\mathbf{a}+\mathrm{q}(\mathrm{q}-\mathrm{p})\mathbf{d}]=2\mathrm{q}[\mathbf{a}+(\mathrm{q}-\mathrm{p})\mathbf{d}]=2\mathrm{q}\alpha. \tag{35.1a}$$

The breeding values of A_1A_2 and A_2A_2, $\mathbf{A}_{12}$ and $\mathbf{A}_{22}$, respectively, can be evaluated in a similar manner. When A_1A_2 is mated to A_1A_2, the offspring are in the 1: 2: 1 ratio. Then

$$(1/4)(\mathbf{a}-\bar{\mathrm{M}})+(1/2)(\mathbf{d}-\bar{\mathrm{M}})+(1/4)(-\mathbf{a}-\bar{\mathrm{M}})$$
$$=(1/4)(\mathbf{a}+2\mathbf{d}-\mathbf{a})-\bar{\mathrm{M}}=(1/2)(\mathbf{d})-\bar{\mathrm{M}}.$$

The breeding value for A_1A_2 is

$$\mathbf{A}_{12}=2[(1/2)\mathrm{p}^2(\mathbf{a}+\mathbf{d})+(1/2)2\mathrm{pq}\mathbf{d}+(1/2)\mathrm{q}^2(\mathbf{d}-\mathbf{a})-\bar{\mathrm{M}}]$$
$$=\mathrm{p}^2\mathbf{a}+\mathrm{p}^2\mathbf{d}+2\mathrm{pq}\mathbf{d}+\mathrm{q}^2\mathbf{d}-\mathrm{q}^2\mathbf{a}-2(\mathrm{p}-\mathrm{q})\mathbf{a}-4\mathrm{pq}\mathbf{d}.$$

The coefficient of **a** in the $\mathbf{A}_{12}$ equation is

$$\mathrm{p}^2-\mathrm{q}^2-2\mathrm{p}+2\mathrm{q}=(\mathrm{p}-\mathrm{q})(\mathrm{p}+\mathrm{q})-2(\mathrm{p}-\mathrm{q})$$
$$=-(\mathrm{p}-\mathrm{q})=(\mathrm{q}-\mathrm{p});$$

that of **d** is

$$\mathrm{p}^2+2\mathrm{pq}+\mathrm{q}^2-4\mathrm{pq}=(\mathrm{p}-\mathrm{q})^2=(\mathrm{q}-\mathrm{p})^2;$$

so that

$$\mathbf{A}_{12}=(\mathrm{q}-\mathrm{p})\mathbf{a}+(\mathrm{q}-\mathrm{p})^2\mathbf{d}=(\mathrm{q}-\mathrm{p})[\mathbf{a}+(\mathrm{q}-\mathrm{p})\mathbf{d}]$$
$$=(\mathrm{q}-\mathrm{p})\alpha. \tag{35.1b}$$

The breeding value of A_2A_2 is

$$\mathbf{A}_{22}=2[\mathrm{p}^2\mathbf{d}+\mathrm{pq}(\mathbf{d}-\mathbf{a})+\mathrm{q}^2(-\mathbf{a})-\bar{\mathrm{M}}]$$
$$=2[\mathrm{p}^2\mathbf{d}+\mathrm{pq}\mathbf{d}-\mathrm{pq}\mathbf{a}-\mathrm{q}^2\mathbf{a}-(\mathrm{p}-\mathrm{q})\mathbf{a}-2\mathrm{pq}\mathbf{d}]$$
$$=2[-\mathrm{p}\mathbf{a}-\mathrm{p}(\mathrm{q}-\mathrm{p})\mathbf{d}]=-2\mathrm{p}\alpha. \tag{35.1c}$$

The value of the A_1 allele is $\mathrm{q}\alpha$; that of the A_2 allele, $-\mathrm{p}\alpha$. Thus the breeding value of each genotype is the sum of the values of the genes in that genotype. That is, the breeding value of A_1A_2, for example, is

$$\mathbf{A}_{12}=(\mathrm{q}-\mathrm{p})\alpha=\mathrm{q}\alpha-\mathrm{p}\alpha=\mathrm{M}(A_1)+\mathrm{M}(A_2).$$

Thus, breeding values are *additive*.

Breeding values are also *linear*. The difference between the breeding value of A_1A_2 and that of A_2A_2 is

$$\mathbf{A}_{12}-\mathbf{A}_{22}=(\mathrm{q}-\mathrm{p})\alpha-(-2\mathrm{p}\alpha)=(\mathrm{q}-\mathrm{p}+2\mathrm{p})\alpha=\alpha.$$

The difference between $\mathbf{A}_{11}$ and $\mathbf{A}_{12}$ is also α:

$$\mathbf{A}_{11}-\mathbf{A}_{12}=2\mathrm{q}\alpha-(\mathrm{q}-\mathrm{p})\alpha=(2\mathrm{q}-\mathrm{q}+\mathrm{p})\alpha=\alpha.$$

Therefore, the breeding values, if plotted against genotypes, will fall along a straight line whose slope is α.

Referring again to our fictitious example of the effect of the head spot locus on body weight in mice, we can calculate the breeding values of the *HsHs*, *Hshs* and *hshs* genotypes, respectively, as

$$\mathbf{A}_{11}=2q\alpha=2(0.4)(0.0038)=0.00304,$$
$$\mathbf{A}_{12}=(q-p)\alpha=(-0.2)(0.0038)=-0.00076,$$
$$\mathbf{A}_{22}=-2p\alpha=-2(0.6)(0.0038)=-0.00456,$$

taking $\alpha=0.0038$ as determined previously. These values are the appropriate functions of the values M(*Hs*) and M(*hs*). They are also linear, because

$$0.00304-(-0.00076)=0.0038,$$
$$-0.00076-(-0.00456)=0.0038;$$

the difference in each case is equal to α.

In Fig. 35.1, the genotypes are displayed at equal intervals along the X-axis, while the Y-axis represents the scale of values. The breeding values in this graph fall upon a straight line, whose slope is α. We shall have more to say about the linearity of breeding values after we have considered dominance deviations in the next lecture.

We still cannot ordinarily identify individual genotypes, and so cannot directly evaluate breeding values for a single locus as derived above. We can, how-

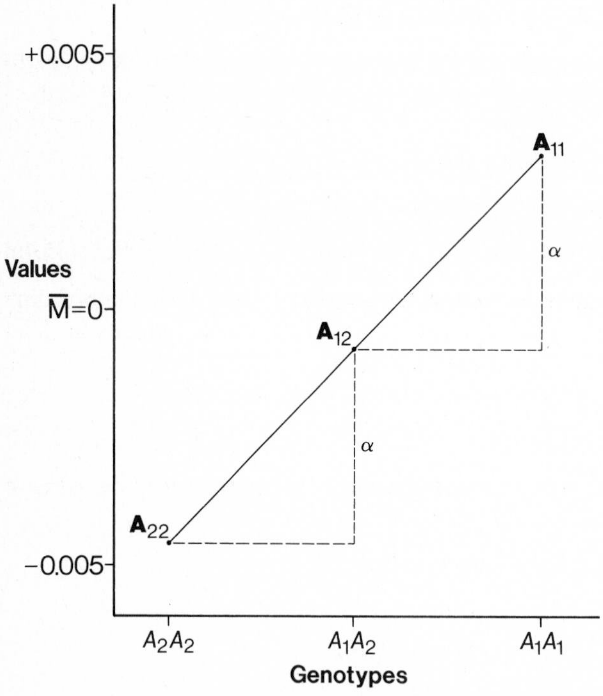

Fig. 35.1. The linear nature of breeding values. The difference between $\mathbf{A}_{11}$ and $\mathbf{A}_{12}$ is the same as that between $\mathbf{A}_{12}$ and $\mathbf{A}_{22}$. Thus, the three $\mathbf{A}$ values fall on a straight line

ever, estimate the breeding value of an individual. If the individual is mated to a random sample of mates, we can calculate the average value of his offspring, and double this to give an estimate of the breeding value of the individual, across all loci.

Just such a process is employed in "proving" dairy bulls for inclusion in an artificial insemination stud. Young bull candidates for the stud are chosen on the basis of the performance of their dams and their sisters. Semen from each candidate is used to inseminate a number of cows; heifer calves are raised, and their milk production measured. Average records for the daughters of each candidate bull are compared. These daughter averages constitute, in fact, estimates of half the breeding values of the candidate bulls, assuming that each has been mated to a random sample of cows. Since they are compared inter alia, multiplication by 2 is not necessary. The elaborate calculations that are applied to the daughter records are basically intended to make the "proofs" of the different bulls more fairly comparable, by removing the effects of differences in environment between herds in which the different bulls' daughters make their records, and of differences in genetic quality between the females to which the different bulls are mated.

Lecture 35 Exercise

Calculate the breeding values of A_1A_1, A_1A_2, and A_2A_2 genotypes for the 27 combinations of parameters given below.

p	a	d
0.9	2	1
0.5	4	0
0.1	6	−1

Lecture 36 Dominance Deviations

The breeding value of a genotype is the sum of the values of the genes in that genotype. Patently, the breeding value will not always be equal to the genotypic value; there may be a difference between $\mathbf{G}_{ii*}$ and $\mathbf{A}_{ii*}$ for genotype A_iA_{i*}. We shall explore the nature of this difference, $\mathbf{D}_{ii*}$.

The values of the $\mathbf{D}_{ii*}$ can be easily found, because

$$\mathbf{D}_{ii*} = \mathbf{G}_{ii*} - \mathbf{A}_{ii*}.$$

The $\mathbf{A}_{ii*}$ are measured on the $\bar{M}$ scale; the $\mathbf{G}_{ii*}$ must also be measured on that scale. Then

$$\begin{aligned}
\mathbf{D}_{11} &= \mathbf{a} - \bar{M} - 2q\alpha \\
&= \mathbf{a} - (p-q)\mathbf{a} - 2pq\mathbf{d} - 2q\mathbf{a} - 2q(q-p)\mathbf{d} \\
&= (1-p+q-2q)\mathbf{a} - (2pq+2q^2-2pq)\mathbf{d} \\
&= -2q^2\mathbf{d}; \\
\mathbf{D}_{12} &= \mathbf{d} - \bar{M} - (q-p)\alpha \\
&= (-p+q-q+p)\mathbf{a} + (1-2pq-q^2+2pq-p^2)\mathbf{d} \\
&= 2pq\mathbf{d}; \\
\mathbf{D}_{22} &= -\mathbf{a} - \bar{M} - (-2p\alpha) \\
&= -(1+p-q-2p)\mathbf{a} - (2pq-2pq+2p^2)\mathbf{d} \\
&= -2p^2\mathbf{d}.
\end{aligned} \tag{36.1}$$

The sum, weighted by the genotypic frequencies, of the $\mathbf{D}_{ii*}$, is zero:

$$p^2\mathbf{D}_{11} + 2pq\mathbf{D}_{12} + q^2\mathbf{D}_{22} = -2p^2q^2\mathbf{d} + 4p^2q^2\mathbf{d} - 2p^2q^2\mathbf{d} = 0.$$

Each of the $\mathbf{D}_{ii*}$ values depends only on $\mathbf{d}$, and is independent of $\mathbf{a}$. If $\mathbf{d}$ were 0, each $\mathbf{D}_{ii*}$ would be 0, and each genotypic value would be equal to the breeding value of that genotype. As we have seen, $\mathbf{d}$ represents the deviation of the heterozygote value from the midpoint between the two homozygotes, and thus measures a quantitative concept of dominance. When $\mathbf{d} = 0$, there is no dominance; the $\mathbf{D}_{ii*}$ are also all zero. Quite properly, then, the $\mathbf{D}_{ii*}$ are referred to as dominance deviations.

The dominance deviation of a genotype represents the difference between the value of the whole genotype and the sum of the values of its genic parts. As such, the dominance deviation is statistically an interaction effect, resulting from the interaction of alleles at the same locus.

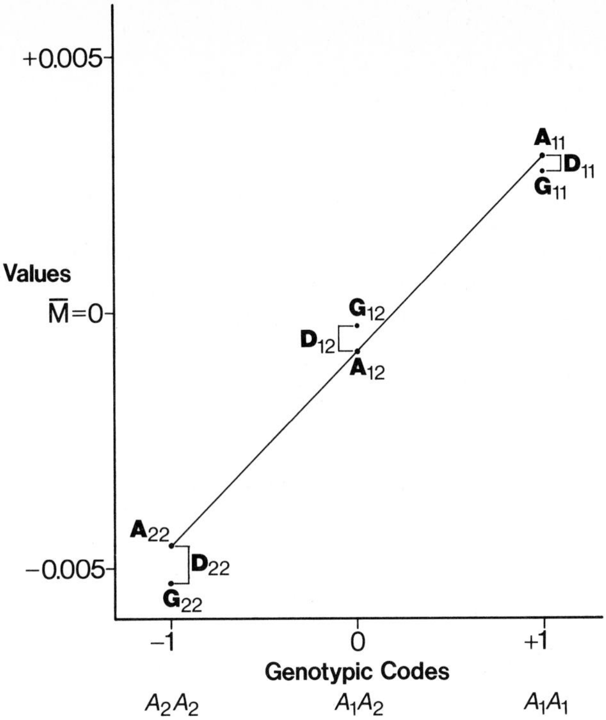

Fig. 36.1. The regression of genotypic value on genotype. The regression line passes through the three points, $\mathbf{A}_{11}$, $\mathbf{A}_{12}$, and $\mathbf{A}_{22}$. Its slope is α; it is, in fact, identical to the line depicted in Fig. 35.1. The vertical distances between the breeding values and the genotypic values are the dominance deviations

We have seen that the breeding values are additive and linear. When $\mathbf{d}=0$, so that each genotypic value equals the corresponding breeding value, it is obvious that additivity and linearity must also apply to the genotypic values, as in Fig. 35.1. In the presence of dominance, however, it is more difficult to identify the additive, linear component in the genotypic values. A straight line obviously cannot be drawn connecting the three genotypic values. (See Fig. 36.1.) We must resort to regression methods to demonstrate the linear component of genotypic values in the presence of dominance.

In order to evaluate the regression of genotypic value on genotype, we must give the genotypes some numerical code value, X_{ii*}, so that we can carry out the arithmetic of the regression calculations. Any equal interval coding will serve. We could, for example, use the number of A_1 alleles in the genotype, coding A_1A_1, A_1A_2, and A_2A_2 as 2, 1, and 0, respectively. Arithmetic calculations will be slightly simpler, however, if we subtract 1 from each of these values, so that our codes are $X_{11}=+1$, $X_{12}=0$, and $X_{22}=-1$. Table 36.1 shows the genotypes with their respective frequencies, codes, and genotypic values, on the $\bar{M}$ scale. Remember that the genotypic codes are not values. X is simply a variable introduced to permit the mechanics of the regression process. From Table 36.1, the value of

Table 36.1. Information for calculation of the regression

Genotype	Frequency	Coding variable = X	Genotypic value = G
A_1A_1	p^2	1	$\mathbf{a}-\bar{M}$
A_1A_2	2pq	0	$\mathbf{d}-\bar{M}$
A_2A_2	q^2	−1	$-\mathbf{a}-\bar{M}$

$b_{\mathbf{G}X}$ can be determined:

$$\Sigma f_{ii*}X_{ii*}=p^2(1)+2pq(0)+q^2(-1)=p^2-q^2=p-q;$$

$$\Sigma f_{ii*}X_{ii*}{}^2=p^2(1)^2+2pq(0)^2+q^2(-1)^2=p^2+q^2;$$

$$\begin{aligned}SSq(X)&=p^2+q^2-(p-q)^2=p^2+q^2-p^2+2pq-q^2\\&=2pq;\end{aligned}$$

$$\begin{aligned}\Sigma f_{ii*}\mathbf{G}_{ii*}&=p^2(\mathbf{a}-\bar{M})+2pq(\mathbf{d}-\bar{M})+q^2(-\mathbf{a}-\bar{M})\\&=(p-q)\mathbf{a}+2pq\mathbf{d}-\bar{M}=0;\end{aligned}$$

$$\begin{aligned}\Sigma(f_{ii*}X_{ii*}\mathbf{G}_{ii*})&=p^2(1)(\mathbf{a}-\bar{M})+2pq(0)(\mathbf{d}-\bar{M})+q^2(-1)(-\mathbf{a}-\bar{M})\\&=(p^2+q^2)\mathbf{a}+(q^2-p^2)\bar{M}\\&=(p^2+q^2)\mathbf{a}+(q-p)(p-q)\mathbf{a}+(q-p)2pq\mathbf{d}\\&=(p^2+q^2-p^2+2pq-q^2)\mathbf{a}+2pq(q-p)\mathbf{d}\\&=2pq\mathbf{a}+2pq(q-p)\mathbf{d}\\&=2pq[\mathbf{a}+(q-p)\mathbf{d}];\end{aligned}$$

$$[\Sigma(f_{ii*}X_{ii*})][(f_{ii*}\mathbf{G}_{ii*}]=(p-q)(0)=0;$$

$$SCP(X\mathbf{G})=2pq[\mathbf{a}+(q-p)\mathbf{d}]-0=2pq[\mathbf{a}+(q-p)\mathbf{d}].$$

Therefore,

$$\begin{aligned}b_{\mathbf{G}X}&=SCP(X\mathbf{G})/SSq(X)=2pq[\mathbf{a}+(q-p)\mathbf{d}]/2pq\\&=\mathbf{a}+(q-p)\mathbf{d}=\alpha.\end{aligned}$$

The slope of the regression line is α. Increasing X by 1 unit will increase **G** by α units, on the average. This agrees with our evaluation of genic values by substitution, since increasing X by one unit is equivalent to substituting an A_1 for an A_2 gene. The slope of the regression line will be equal to α, however, only if we assume random mating and a Hardy-Weinberg genotypic ratio.

By definition, the regression line passes through the point $\bar{X}$, $\bar{\mathbf{G}}$:

$$\bar{X}=\Sigma f_{ii*}X_{ii*}/\Sigma f_{ii*}=p-q;$$

$$\bar{\mathbf{G}}=\Sigma f_{ii*}\mathbf{G}_{ii*}/\Sigma f_{ii*}=0.$$

We can therefore draw the regression line shown in Fig. 36.1.

Table 36.2. G, A, and D values for the head spot example

Genotype	G_{ii*}	A_{ii*}	D_{ii*}
A_1A_1	0.00272	0.00304	−0.00032
A_1A_2	−0.00028	−0.00076	0.00048
A_2A_2	−0.00528	−0.00456	−0.00072

The $\hat{\mathbf{G}}_{ii*}$ are the values of the genotypes predicted by the regression, based on the formula

$$\hat{\mathbf{G}}_{ii*} - \bar{\mathbf{G}} = \alpha(X_{ii*} - \bar{X}).$$

These values are therefore

$$\hat{\mathbf{G}}_{11} = \alpha[1-(p-q)] = 2q\alpha = \mathbf{A}_{11};$$

$$\hat{\mathbf{G}}_{12} = \alpha[0-(p-q)] = (q-p)\alpha = \mathbf{A}_{12};$$

$$\hat{\mathbf{G}}_{22} = \alpha[-1-(p-q)] = -2p\alpha = \mathbf{A}_{22}.$$

The values of the genotypes predicted by regression are their breeding values. These are, of course, linear and additive. It follows that the deviations of the actual values from the regression line are the dominance deviations.

Referring once more to our fictitious head spot example, we had found that $\mathbf{A}=0.004$, $\mathbf{d}=0.001$. Therefore, on the $\tilde{y}$ scale, the genotypic values of *HsHs*, *Hshs*, and *hshs* were 0.004, 0.001 and −0.004, respectively. $\bar{M}=0.00128$, if we assume $p=0.6$ for *Hs*, $q=0.4$ for *hs*. Therefore, $\mathbf{G}_{11}=0.00272$, $\mathbf{G}_{12}=-0.00028$, $\mathbf{G}_{22}=-0.00528$, on the $\bar{M}$ scale.

Allelic effects are M(*Hs*) = 0.00152, M(*hs*) = −0.00228, so that $\mathbf{A}_{11}=0.00304$, $\mathbf{A}_{12}=-0.00076$, and $\mathbf{A}_{22}=-0.00456$. Subtracting these breeding values from the genotypic values yields the dominance deviations, $\mathbf{D}_{11}=-0.00032$, $\mathbf{D}_{12}=0.00048$, and $\mathbf{D}_{22}=-0.00072$. The weighted sum of the $\mathbf{D}_{ii*}$ is 0. These results are summarized in Table 36.2, and are presented on the graph in Fig. 36.1.

The regression line runs through the $\mathbf{A}_{ii*}$ values; the deviation of the $\mathbf{G}_{ii*}$ from the line are the $\mathbf{D}_{ii*}$. $\mathbf{D}_{12}$ is positive, while $\mathbf{D}_{11}$ and $\mathbf{D}_{22}$ are negative. It is necessary that $\mathbf{D}_{12}$ have the opposite sign to the other two $\mathbf{D}_{ii*}$ values in order that the weighted sum of the $\mathbf{D}_{ii*}$ be zero. (The regression calculation is a least squares procedure.) $\mathbf{D}_{12}$ is positive in our example because $\mathbf{G}_{12}$ is closer to $\mathbf{G}_{11}$ than to $\mathbf{G}_{22}$. Had $\mathbf{G}_{12}$ been less than the midpoint between $\mathbf{G}_{11}$ and $\mathbf{G}_{22}$, $\mathbf{D}_{12}$ would have been negative and the other two positive.

$\mathbf{D}_{22}$ is greater in absolute value than $\mathbf{D}_{11}$. In order that the weighted sum of the $\mathbf{D}_{ii*}$ should equal 0,

$$p^2\mathbf{D}_{11} = q^2\mathbf{D}_{22} = (1/2)(2pq\mathbf{D}_{12}).$$

Therefore, the less frequent homozygote will have a larger dominance deviation than the more frequent homozygote.

Lecture 36 Exercise

Calculate the $\mathbf{D}_{ii*}$ values for the parameters in the Lecture 35 exercise:

p	a	d
0.9	2	1
0.5	4	0
0.1	6	−1

Lecture 37 Multiple Loci

The model we have constructed in the last several lectures was based upon a single locus. At one locus, we have concluded that the value of genotype A_iA_{i*} is the sum of its breeding value plus a deviation due to the effect of dominance. In symbols,

$$\mathbf{G}_{ii*} = \mathbf{A}_{ii*} + \mathbf{D}_{ii*}.$$

The breeding value is the sum of the values of the genes A_i and A_{i*}; the dominance deviation is an effect of the interaction between these two genes.

Metric traits are, however, characteristically polygenic, inherited by the effects of a number of loci. We must therefore expand our model to account for multiple loci before we can use it to explain quantitative inheritance.

We simply define the multilocus breeding value as the sum of the breeding values at each locus involved in the inheritance of a given trait. Suppose, for example, that individual ij has genotype

$$W_1W_1X_1X_2Y_1Y_2Z_2Z_2$$

at the W, X, Y and Z loci, and that these four loci, and only these, affect some metric trait. The breeding value of individual ij for that trait would be:

$$\begin{aligned} \mathrm{A_i} &= \mathbf{A}_{11(\mathrm{W})} + \mathbf{A}_{12(\mathrm{X})} + \mathbf{A}_{12(\mathrm{Y})} + \mathbf{A}_{22(\mathrm{Z})} \\ &= 2q_W\alpha_W + (q_X - p_X)\alpha_X + (q_Y - p_Y)\alpha_Y - 2p_Z\alpha_Z. \end{aligned} \quad (37.1)$$

(An **A**, in boldface type, will be used to designate the additive effects, or breeding values, at a single locus, while A represents the sum of breeding effects across all loci. A similar distinction will be made between **G** and G, and between **D** and D.)

The multiple locus breeding value is the sum, across all loci, of the breeding values at individual loci. These in turn are the sums of the genic effects within a locus. Thus, the multiple locus breeding value of a given genotype is the grand total of the genic effects of all the genes in that genotype, across all loci. Breeding values are additive, not only within, but across loci; hence we refer to them as additive effects.

Dominance deviations arise from interactions between alleles at the same locus. The multiple locus dominance deviation is defined as the sum of the individual locus dominance effects:

$$\mathrm{D_i} = \mathbf{D}_{11(\mathrm{W})} + \mathbf{D}_{12(\mathrm{X})} + \mathbf{D}_{12(\mathrm{Y})} + \mathbf{D}_{22(\mathrm{Z})}. \quad (37.2)$$

The **D** value at each locus is, of course, defined in terms of the gene frequencies and **d** for that locus, e.g., $\mathbf{D}_{11(\mathrm{W})} = -2q_W\mathbf{d}_W$, etc.

$\mathrm{G_i}$ is the total effect of the multilocus genotype i. This total may not be equivalent to the sum of its additive plus dominance components.

We define the difference as I:

$$G_i = A_i + D_i + I_i. \tag{37.3}$$

Since A_i and D_i are the sums of individual locus effects, I_i is, statistically, an interaction between loci. Interlocus interactions occur in qualitative traits, the classic example being albinism, which occurs in many species. Suppose we have a stock of mice segregating at the brown pigment locus. We expect *BB* and *Bb* mice to have black pigment, *bb* mice to have brown. If the stock is also segregating for the albino gene, *bb* mice do, indeed, have brown pigment, and the other two genotypes black pigment, but only if their genotypes at the albino locus are either *CC* or *Cc*. *cc* mice have no pigment, black or brown, regardless of their genotypes at the brown pigment locus. The albino genotype masks the effects of the genotype at the brown locus; it is epistatic to the brown locus genotype, and to the genotype at any other locus controlling quality, quantity, and distribution of pigmentation. The other color loci are hypostatic to the albino genotype.

Thus, in a qualitative trait, the effect of an epistatic genotype is to mask the effect of genotypes at other loci which are hypostatic to it, much as a dominant allele masks the effects of alleles at the same locus which are recessive to it. Epistasis is a sort of interlocus dominance; like dominance, it is an interaction between genes, but genes at different loci.

We extended the qualitative concept of dominance to metric inheritance, defining it as a statistical interaction causing the genotypic value at a single locus to vary from the sum of the effects of the genes at that locus. The effect of interaction between genes and/or genotypes at different loci will be to cause the genotypic value across all loci to vary from the sum of the additive and dominance effects across all loci, as recognized by Fisher (1918). This interaction is the epistatic interaction, I_i.

If it were possible to identify genotypes for a quantitative trait, and thus to separate the effects of breeding values, dominance deviations, and epistatic interactions, we would find that the latter were classifiable on two planes. One plane would involve the number of loci interacting; some interactions would involve two loci, others three, and so on up to the total number of loci contributing to the inheritance of that particular trait. The other plane would involve the types of intralocus effects involved in the interaction. At two loci, we could have interactions between individual genes at both, or between gene pairs at both, or between individual genes at one locus and gene pairs at the other. Thus, not one, but a number of epistatic interactions may occur with regard to the inheritance of a given trait. In our model, we usually lump all of these together as a single factor, I.

In defining the general quantitative model,

$$P_{ij} = \mu m + G_i + E_{ij},$$

P_{ij} was the phenotypic value for some metric trait of the j-th individual with genotype i, and μm was a general mean over all individuals in the population. G_i was the deviation from μm caused by the i-th genotype (the joint effect of all gene loci involved in the inheritance of the trait), and E_{ij} was the deviation of P_{ij} from $\mu m + G_i$ due to environmental effects on individual ij.

We have now defined additive, dominance, and epistatic components of G, setting

$$G_i = A_i + D_i + I_i, \tag{37.4}$$

for the i-th genotype. A_i is the sum of the additive effects at each single locus, and these single locus effects are defined as deviations from the mean effect of that locus, $\bar{M}$. Thus, A_i is a deviation from the sum of the $\bar{M}$ for all loci. That sum is part of μm, along with any nongenetic effects which are common to all individuals in the population, when we substitute the genetic components into the general model:

$$P_{ij} = \mu m + A_i + D_i + I_i + E_{ij}. \tag{37.5}$$

Lecture 37 Exercise

Suppose that we have two loci, U and V, each segregating for a pair of alleles in a Hardy-Weinberg population of a diploid organims. Each locus has codominant effects on a qualitative trait, so that we can identify each genotype separately by its phenotype.

We are interested in the effects these loci may have on a metric trait. We measure the metric trait on n individuals from each of the 9 genotypes, n being a large number. The average value for each genotype, on the $\bar{M}$ scale, is shown in the table:

	U_1U_1	U_1U_2	U_2U_2
V_1V_1	6	3	−3
V_1V_2	3	3	−3
V_2V_2	0	−3	−6

Using these data, we can estimate the various parameters for each genotype at the U and V loci. We find the following results for the sum of the **A** and **D** parameters at the two loci:

	U locus	V locus
$\mathbf{A}_{11}+\mathbf{D}_{11}$	3	2
$\mathbf{A}_{12}+\mathbf{D}_{12}$	1	1
$\mathbf{A}_{22}+\mathbf{D}_{22}$	−4	−3

Evaluate epistatic interactions for each of the 9 genotypes.

Lecture 38 Causal Components of Variance

We have seen that we can equate the phenotypic value of an individual to the sum of the effects of its genotype and environment:

$$P_{ij} = \mu m + G_i + E_{ij}. \qquad (38.1)$$

Although we cannot directly evaluate genotypic effects from Eq. (38.1), this equation is a linear model, and a powerful statistical tool, the analysis of variance, exists which can be applied to linear models. The basic purpose of the analysis of variance is to partition total variance into components associated with the factors of the model. In the case of the model represented by Eq. (38.1), the total variance among phenotypes will be partitioned into components that can be associated with genotypic and with environmental effects:

$$\sigma_P{}^2 = \sigma_G{}^2 + \sigma_E{}^2. \qquad (38.2)$$

The above expression of the phenotypic variance as the sum of genotypic and environmental variances ignores any covariance between genotype and environment. We assume that genotype and environment are uncorrelated; the covariance of the two factors is 0. Therefore, there is no covariance term in the above equation.

The genotypic variance, $\sigma_G{}^2$, is the variance due to the total effect of genotypes. We have, in previous lectures, subdivided the genotypic effect, G_i, into additive, dominance and epistatic interaction effects. In a similar manner, we can subdivide the genotypic variance into components due to additive, dominance and epistatic interaction effects:

$$\sigma_G{}^2 = \sigma_A{}^2 + \sigma_D{}^2 + \sigma_I{}^2.$$

We have defined these effects so that they are uncorrelated, the covariances between them being 0, because we measure A, D, and I as deviations from $\bar{M}$, $\bar{M}+A$, and $\bar{M}+A+D$, respectively.

With this partition of the genotypic variance, we can write

$$\sigma_P{}^2 = \sigma_G{}^2 + \sigma_E{}^2 = \sigma_A{}^2 + \sigma_D{}^2 + \sigma_I{}^2 + \sigma_E{}^2. \qquad (38.3)$$

We can calculate the theoretical variances due to breeding values at a single locus. The sum of the breeding values at individual loci is, as we have seen, 0, and therefore their average is zero. But then,

$$\begin{aligned} \mathrm{Var}(\mathbf{A}_{ii*}) &= p^2(2q\alpha)^2 + 2pq(q-p)^2\alpha^2 + q^2(-2p\alpha)^2 \\ &= 4p^2q^2\alpha^2 + 2pq(q^2 - 2pq + p^2)\alpha^2 + 4p^2q^2\alpha^2 \\ &= 2pq\alpha^2(2pq + q^2 - 2pq + p^2 + 2pq) = 2pq\alpha^2(p+q)^2 \\ &= 2pq\alpha^2. \end{aligned} \qquad (38.4a)$$

Similarly, the average dominance deviation is 0, and

$$\begin{aligned}\mathrm{Var}(\mathbf{D}_{ii*}) &= p^2(-2q^2\mathbf{d})^2 + 2pq(2pq\mathbf{d})^2 + q^2(-2p^2\mathbf{d})^2 \\ &= 4p^2q^4\mathbf{d}^2 + 8p^3q^3\mathbf{d}^2 + 4p^4q^2\mathbf{d}^2 \\ &= 4p^2q^2\mathbf{d}^2(q^2 + 2pq + p^2) = 4p^2q^2\mathbf{d}^2. \end{aligned} \tag{38.4b}$$

The variance of genotypic values at a single locus can be calculated as well:

$$\begin{aligned}\mathrm{Var}(\mathbf{G}_{ii*}) &= p^2(\mathbf{a} - \bar{M})^2 + 2pq(\mathbf{d} - \bar{M})^2 + q^2(-\mathbf{a} - \bar{M})^2 \\ &= p^2(\mathbf{a}^2 - 2\mathbf{a}\bar{M} + \bar{M}^2) + 2pq(\mathbf{d}^2 - 2\mathbf{d}\bar{M} + \bar{M}^2) + q^2(\mathbf{a}^2 + 2\mathbf{a}\bar{M} + \bar{M}^2) \\ &= (p^2 + q^2)\mathbf{a}^2 + 2pq\mathbf{d}^2 + 2(q^2 - p^2)\mathbf{a}\bar{M} - 4pq\mathbf{d}\bar{M} + \bar{M}^2. \end{aligned} \tag{38.5}$$

$\mathrm{Var}(\mathbf{G}_{ii*})$ is expressed in terms of $\mathbf{a}^2$, $\mathbf{d}^2$, $\mathbf{a}\bar{M}$, $\mathbf{d}\bar{M}$, and $\bar{M}^2$, where

$$\bar{M} = (p - q)\mathbf{a} + 2pq\mathbf{d}.$$

Then

$$\mathbf{a}\bar{M} = (p - q)\mathbf{a}^2 + 2pq\mathbf{ad},$$

$$\mathbf{d}\bar{M} = (p - q)\mathbf{ad} + 2pq\mathbf{d}^2,$$

and

$$\bar{M}^2 = (p - q)^2\mathbf{a}^2 + 4pq(p - q)\mathbf{ad} + 4p^2q^2\mathbf{d}^2.$$

Substituting these into Eq. (38.5), we find

$$\begin{aligned}\mathrm{Var}(\mathbf{G}_{ii*}) = {} & (p^2 + q^2)\mathbf{a}^2 + 2pq\mathbf{d}^2 - 2(q - p)^2\mathbf{a}^2 + 4pq(q - p)\mathbf{ad} - 4pq(p - q)\mathbf{ad} \\ & - 8p^2q^2\mathbf{d}^2 + (p - q)^2\mathbf{a}^2 + 4pq(p - q)\mathbf{ad} + 4p^2q^2\mathbf{d}^2. \end{aligned}$$

In the above equation, the coefficients of $\mathbf{a}^2$ are

$$(p^2 + q^2) - 2(q - p)^2 + (p - q)^2 = p^2 + q^2 - p^2 + 2pq - q^2 = 2pq.$$

The coefficients of **ad** are

$$4pq(q - p) + 4pq(p - q) - 4pq(p - q) = 4pq(q - p);$$

and those of $\mathbf{d}^2$ are

$$2pq - 8p^2q^2 + 4p^2q^2 = 2pq(1 - 2pq) = 2pq(p^2 + q^2).$$

The variance of the genotypic values therefore reduces to

$$\begin{aligned}\mathrm{Var}(\mathbf{G}_{ii*}) &= 2pq\mathbf{a}^2 + 4pq(q - p)\mathbf{ad} + 2pq(p^2 + q^2)\mathbf{d}^2 \\ &= 2pq[\mathbf{a}^2 + 2(q - p)\mathbf{ad} + (p^2 + q^2)\mathbf{d}^2]. \end{aligned}$$

Note, however, that

$$(q - p)^2 = (q^2 + p^2 - 2pq).$$

Therefore,

$$(p^2 + q^2) = (q - p)^2 + 2pq,$$

and we can write

$$\begin{aligned}\mathrm{Var}(\mathbf{G}_{ii*}) &= 2pq[\mathbf{a}^2+2(q-p)\mathbf{ad}+(q-p)^2\mathbf{d}^2]+(2pq)^2\mathbf{d}^2\\ &= 2pq[\mathbf{a}+(q-p)\mathbf{d}]^2+4p^2q^2\mathbf{d}^2 = 2pq\alpha^2+4p^2q^2\mathbf{d}^2\\ &= \mathrm{Var}(\mathbf{A}_{ii*})+\mathrm{Var}(\mathbf{D}_{ii*})\end{aligned} \tag{38.6}$$

Thus, we see that at a single locus the genotypic variance is indeed the sum of the additive and dominance variances. (At a single locus, of course, epistatic interactions do not enter into the genotypic variance, as these involve interactions among two or more loci.)

We noted previously that, in the multiple locus model, the effect of the i-th genotype (across loci) was the sum of the additive effects at each locus, plus the sum of the dominance effects at each locus, plus the epistatic interaction effects among loci. Thus,

$$G_i = A_i + D_i + I_i,$$

with G_i standing for the total effects of genotypes across all loci, and so forth. Because, for loci X, Y, Z, ...,

$$A_i = \mathbf{A}_{ii*(X)} + \mathbf{A}_{ii*(Y)} + \mathbf{A}_{ii*(Z)} + \ldots,$$

then

$$\sigma_A{}^2 = \mathrm{Var}_X(\mathbf{A}_{ii*}) + \mathrm{Var}_Y(\mathbf{A}_{ii*}) + \mathrm{Var}_Z(\mathbf{A}_{ii*}) + \ldots.$$

Similarly, $\sigma_D{}^2$ is the sum of the $\mathrm{Var}(\mathbf{D}_{ii*})$ at the various loci. Finally,

$$\sigma_I{}^2 = \sigma_{AA}{}^2 + \sigma_{AD}{}^2 + \sigma_{DD}{}^2 + \sigma_{AAA}{}^2 + \ldots, \tag{38.1}$$

where $\sigma_{AA}{}^2$ is the variance due to interactions between single gene effects at two loci, summed across all pairs of loci; $\sigma_{AD}{}^2$ is the variance due to interactions between single gene effects at one locus and gene pair effects at a second, summed over all pairs of loci; and so forth.

(Conventional notation uses the subscript A to symbolize single gene, and D, gene pair, effects in these epistatic variances. Furthermore, $\sigma_{AA}{}^2$ is referred to as "additive by additive" variance, and so forth. Additive effects are the effects of single genes, and dominance effects those of gene pairs, but the implication that $\sigma_{AA}{}^2$ is due to the interaction of additive effects at two loci is not correct.)

Lecture 38 Exercises

1. A quantitative trait in a certain organism is known to be inherited by the action of three gene loci, X, Y and Z. In a given population, the following parameters have been measured for each of the three loci:

	X-locus	Y-locus	Z-locus
p	0.6	0.5	0.4
α	1.2	1.2	0.6
d	0	0.4	−0.8

Calculate $\sigma_A{}^2$ and $\sigma_D{}^2$ for this trait.

2. The phenotypic standard deviation, σ_P, in the above population for the given trait is approximately 1.7320. Two-thirds of the phenotypic variance of the trait is known to be of genotypic origin. Estimate $\sigma_I{}^2$ and $\sigma_E{}^2$.

Lecture 39 Analytical Components

Our theoretical framework for quantitative inheritance is now complete. The phenotypic value for any individual is caused by its genotype and its environment. The genotypic effect can be subdivided into additive effects of genes; dominance deviations, interactions between effects of genes at the same locus; and epistatic interactions between effects of genes at different loci. Individuals differ in all three of these factors, so that the variance of phenotypic values is caused by additive, dominance, epistatic, and environmental variance components; the first three of these, taken together, constitute the genotypic variance. Since they relate to the factors that actually cause variation, they can be termed the causal factors of phenotypic variance.

The analysis of variance will permit us to estimate variance components associated with groups of observations. If we could group together individuals of the same genotype, we could directly analyze for variance components due to the causal factors. But this we cannot do, because we cannot identify genotypes.

We can, however, group together biologically related individuals, and estimate the covariances within such groups. Biologically related individuals tend to have similar genotypes, and at least part of the covariance between such individuals is due to this fact. We can estimate covariances of relatives from data. These covariances are the analytical components of variance, that is, the components which we can estimate from the results of analyses. If we can express the analytical components as functions of the causal, we can estimate causal from analytical components.

We will look in particular at the genetic covariances between parent and offspring, among half sibs, and among full sibs. These are the three closest relationships to be found in a population. More distant relationships occur, and the covariances among these more distant relatives will also be due partly to genotypic resemblances. But these three covariances have the greatest genotypic contributions, and are the ones most likely to be used in the genetic analysis of a metric trait.

Consider first the genetic covariance between parent and offspring. Suppose that we choose one individual of each of the three A locus genotypes, A_1A_1, A_1A_2, and A_2A_2. Each of these individuals is mated to a series of mates chosen at random from a large random mating population; a single offspring is produced from each mating. The number of mates in each series is large enough that the frequency of matings with mates of genotype A_1A_1 is p^2, with A_1A_2 is 2pq, and with A_2A_2 is q^2, for each of the three individuals.

The genotypic values of the parents are therefore $\mathbf{a}-\bar{M}$, $\mathbf{d}-\bar{M}$, and $-\mathbf{a}-\bar{M}$. We can write these values in terms of α and $\mathbf{d}$:

$$\begin{aligned}
\mathbf{a}-\bar{M} &= \mathbf{a}-(p-q)\mathbf{a}-2pq\mathbf{d}=(1-p+q)\mathbf{a}-2pq\mathbf{d}\\
&=2q\mathbf{a}-2pq\mathbf{d}=2q[\mathbf{a}-(p-q+q)\mathbf{d}]\\
&=2q[\mathbf{a}+(q-p)\mathbf{d}-q\mathbf{d}]\\
&=2q\alpha-2q^2\mathbf{d};\\
\mathbf{d}-\bar{M} &= \mathbf{d}-(p-q)\mathbf{a}-2pq\mathbf{d}=(q-p)\mathbf{a}+(1-2pq)\mathbf{d}\\
&=(q-p)\mathbf{a}+(q^2-2pq+p^2+2pq)\mathbf{d}\\
&=(q-p)\mathbf{a}+(q-p)^2\mathbf{d}+2pq\mathbf{d}\\
&=(q-p)[\mathbf{a}+(q-p)\mathbf{d}]+2pq\mathbf{d}=(q-p)+2pq\mathbf{d};\\
-\mathbf{a}-\bar{M} &= -\mathbf{a}-(p-q)\mathbf{a}-2pq\mathbf{d}=-2p\mathbf{a}-2pq\mathbf{d}\\
&=2p[\mathbf{a}+(q-p+p)\mathbf{d}]=-2p[\mathbf{a}+(q-p)\mathbf{d}+p\mathbf{d}]\\
&=-2p\alpha-2p^2\mathbf{d}.
\end{aligned}$$

Each offspring results from a mating with a random mate. Hence, the average value of the offspring from matings of a parent of genotype A_iA_{i*} will be half the breeding value of that genotype. Table 39.1 shows the parental genotypes with their frequencies, genotypic values, and offspring values.

The genetic covariance of parent and offspring is the average of the crossproducts of parent and offspring values, weighted by the parental frequencies. These crossproducts have been included in Table 39.1 for the sake of convenience. The genetic covariance of parent with offspring at a single locus is therefore:

$$\begin{aligned}
\text{Covar(PO)} &= 2p^2q^2\alpha^2-2p^2q^3\alpha\mathbf{d}+pq(q-p)^2\alpha^2\\
&\quad +2p^2q^2(q-p)\alpha\mathbf{d}+2p^2q^2\alpha^2+2p^3q^2\alpha\mathbf{d}\\
&=(2p^2q^2+pq^3-2p^2q^2+p^3q+2p^2q^2)\alpha^2\\
&\quad +(-2p^2q^3+2p^2q^3-2p^3q^2+2p^3q^2)\alpha\mathbf{d}\\
&=(pq^3+2p^2q^2+p^3q)\alpha^2=(pq)(q^2+2pq+p^2)\alpha^2\\
&=pq\alpha^2=\sigma_{\mathbf{A}}{}^2/2. \qquad (39.1)
\end{aligned}$$

Table 39.1. Parent offspring and half sib covariances

Genotype	Frequency	Parental value (P)	Offspring value (O)	P×O	O²
A_1A_1	p^2	$2q(\alpha-q\mathbf{d})$	$q\alpha$	$2q^2\alpha^2-2q^3\alpha\mathbf{d}$	$q^2\alpha^2$
A_1A_2	$2pq$	$(q-p)\alpha+2pq\mathbf{d}$	$(q-p)\alpha/2$	$(q-p)^2\alpha^2/2$ $+pq(q-p)\alpha\mathbf{d}$	$(q-p)^2\alpha^2/4$
A_2A_2	q^2	$-2p(\alpha+p\mathbf{d})$	$-p\alpha$	$2p^2\alpha^2+2p^3\alpha\mathbf{d}$	$p^2\alpha^2$

The genetic covariance of parent and offspring is half of the variance due to additive genetic effects at a single locus. This is true at every locus; therefore, the covariance of parent and offspring across all loci is half of the additive genetic variance across all loci, $\sigma_A{}^2$. For the moment, we will ignore the epistatic variance component.

The covariance between parent and offspring is a function of the additive genetic variance; dominance variance does not enter into this covariance if the parents are unrelated. Additive effects are properties of individual genes. Individual genes are transmitted from parent to offspring, so that a parent and its offspring must carry an identical gene in common. Therefore parent and offspring covary due to additive effects. Dominance effects are properties of genotypes, and, as we have previously noted, if the parents are unrelated, parent and offspring cannot have identical genotypes. Thus dominance variance cannot contribute to the covariance of parent with offspring, if the parent is not related to the other parent of that offspring.

Next let us consider the genetic covariance among half sibs. The model for the parent offspring covariance envisaged one individual of each A locus genotype mated to a series of random mates, each mating producing one offspring. The offspring of the A_iA_{i*} parent therefore constitute a half sib family, and the offspring values in Table 39.1 are half sib family averages.

If we have t treatments, with n observations within each treatment group, the analysis of variance permits us to divide the total variance among all observations into variance within and variance among treatment groups. The within treatment variance is simply the average of the variance within treatment 1, the variance within treatment 2, etc. The among treatment variance is the variance among the means of the treatment groups. In terms of our model for covariances, the variance among the offspring values in Table 39.1 represents the variance among half sib family means.

The prefix co- means "together"; covariance defines the extent to which two entities vary together. The variance among groups means represents the tendency of members of one group to vary together, to covary, as against members of other groups. It therefore represents the average covariance among members of the same group. The groups we are concerned with in our analysis of half sib covarinace are half sib families. The genetic variance among half sib family means measures the genetic covariance within half sib families. This variance is, of course, the sum of the squared offspring values in Table 39.1, each weighted by the corresponding frequency. The squares of the offspring values are presented in Table 39.1.

$$\begin{aligned}\text{Covar(HS)} &= p^2q^2\alpha^2 + pq(q-p)^2\alpha^2/2 + p^2q^2\alpha^2 \\ &= (p^2q^2 + pq^3/2 - p^2q^2 + p^3q/2 + p^2q^2)\alpha^2 \\ &= (pq^3/2 + p^2q^2 + p^3q/2)\alpha^2 \\ &= (pq/2)(q^2 + 2pq + p^2)\alpha^2 \\ &= pq\alpha^2/2 = \sigma_A{}^2/4. \qquad (39.2)\end{aligned}$$

Again the half sib covariance depends only on additive, not on dominance variance. With unrelated parents, half sibs are related to one another only through the common parent, and therefore can have identical genes, but not identical genotypes.

The genetic covariance between half sibs is only half as great as that between parent and offspring. The probability that the common parent will contribute the same gene to each of a pair of half sibs is only one half, whereas one of the parent's genes will always be present in each offspring.

Each offspring has two parents, and receives half its genes from each. Hence, offspring might be expected to have values close to the average value of their two parents. (If inheritance were strictly additive, with no environmental variance, the expected value of the offspring would be exactly equal to the parents' average.) In recognition of this relationship, a midparent value is often calculated, which is simply the mean of the parental values:

$$\bar{P}=(P_m+P_f)/2.$$

To examine the covariance of midparent and offspring values, we imagine a set of random matings, and calculate the average parental and average offspring values for each (see Table 39.2). Midparent and offspring values in Table 39.2 are expressed in terms of **a** and **d**, and not as deviations from $\bar{M}$. Therefore, the midparent offspring genetic covariance must be calculated in the form

$$\mathrm{Covar}(\bar{P}O)=\Sigma f_i\bar{P}_iO_i-\bar{M}^2,$$

because $\bar{M}$ is the mean of both the midparent values and the offspring values. For convenience in calculation, the crossproducts of midparent and offspring values have been provided in Table 39.2.

We can then evaluate

$$\begin{aligned}\Sigma f_i\bar{P}_iO_i &= p^4\mathbf{a}^2+p^3q(\mathbf{a}+\mathbf{d})^2+2p^2q^2\mathbf{d}^2+pq^3(\mathbf{d}-\mathbf{a})^2+q^4\mathbf{a}^2\\ &=(p^4+p^3q+pq^3+q^4)\mathbf{a}^2+(2p^3q-2pq^3)\mathbf{ad}\\ &\quad+(p^3q+2p^2q^2+pq^3)\mathbf{d}^2\\ &=[p^3(p+q)+q^3(p+q)]\mathbf{a}^2+2pq(p^2-q^2)\mathbf{ad}\\ &\quad+pq(p^2+2pq+q^2)\mathbf{d}^2\\ &=(p^3+q^3)\mathbf{a}^2+2pq(p-q)\mathbf{ad}+pq\mathbf{d}^2,\end{aligned}$$

Table 39.2. Midparent offspring and full sib covariances

Mating	Frequency	Midparent values ($\bar{P}$)	Offspring values (O)	$\bar{P}\times O$	O^2
$A_1A_1\times A_1A_1$	p^4	$\mathbf{a}$	$\mathbf{a}$	$\mathbf{a}^2$	$\mathbf{a}^2$
$A_1A_1\times A_1A_2$	$4p^3q$	$(\mathbf{a}+\mathbf{d})/2$	$(\mathbf{a}+\mathbf{d})/2$	$(\mathbf{a}+\mathbf{d})^2/4$	$(\mathbf{a}+\mathbf{d})^2/4$
$A_1A_1\times A_2A_2$	$2p^2q^2$	0	$\mathbf{d}$	0	$\mathbf{d}^2$
$A_1A_2\times A_1A_2$	$4p^2q^2$	$\mathbf{d}$	$\mathbf{d}/2$	$\mathbf{d}^2/2$	$\mathbf{d}^2/4$
$A_1A_2\times A_2A_2$	$4pq^3$	$(\mathbf{d}-\mathbf{a})/2$	$(\mathbf{d}-\mathbf{a})/2$	$(\mathbf{d}-\mathbf{a})^2/4$	$(\mathbf{d}-\mathbf{a})^2/4$
$A_2A_2\times A_2A_2$	q^4	$-\mathbf{a}$	$-\mathbf{a}$	$\mathbf{a}^2$	$\mathbf{a}^2$

so that

$$\begin{aligned}\text{Covar}(\bar{P}O) &= (p^3+q^3-p^2+2pq-q^2)\mathbf{a}^2-2pq(p-q)\mathbf{ad}\\ &\quad +(pq-4p^2q^2)\mathbf{d}^2\\ &=[2pq-p^2(1-p)-q^2(1-q)]\mathbf{a}^2-2pq(p-q)\mathbf{ad}\\ &\quad +pq(1-4pq)\mathbf{d}^2\\ &=pq(2-p-q)\mathbf{a}^2-2pq(p-q)\mathbf{ad}+pq(p-q)^2\mathbf{d}^2\\ &=pq[\mathbf{a}^2-2(p-q)\mathbf{ad}+(p-q)^2\mathbf{d}^2]\\ &=pq[\mathbf{a}+(q-p)\mathbf{d}]^2=pq\alpha^2=\sigma_\mathbf{A}{}^2/2 \qquad (39.3)\end{aligned}$$

The genetic covariance of midparent with offspring is exactly the same as that of parent with offspring. In a sense, Covar($\bar{P}$O), the covariance of midparent with offspring, is the average of the covariances of each of the two parents with the offspring; and since these two Covar(PO) are each $\sigma_\mathbf{A}{}^2/2$, their average must be the same. Dominance variance does not contribute to the midparent offspring covariance, again, because the parents contribute only genes, not genotypes, to the offspring. The offspring cannot have a genotype identical to the genotype of either parent, unless the parents are related.

The offspring of any one mating are full sibs. If the matings imagined to occur to provide data for the midparent offspring covariance calculations above each produce more than one offspring, offspring from the same mating are full sibs, and the offspring values in Table 39.2 are the means of full sib families. The covariance within full sib families is equivalent to the variance among full sib family means, so that this covariance is

$$\text{Covar(FS)}=\Sigma f_i O_i{}^2.$$

The squared offspring values are provided in Table 39.2.

It can be seen that the squares of the offspring values are equal to the crossproducts of the midparent with offspring values, except in the cases of matings $A_1A_1 \times A_2A_2$ and $A_1A_2 \times A_1A_2$. In the first of these, the crossproduct is 0, while the squared offspring value is $\mathbf{d}^2$. In the second, the crossproduct is $\mathbf{d}^2/2$ while the square is $\mathbf{d}^2/4$. Therefore, we can calculate Covar(FS) as

$$\text{Covar(FS)}=\text{Covar}(\bar{P}O)-4p^2q^2\mathbf{d}^2/2+4p^2q^2\mathbf{d}^2/4+2p^2q^2\mathbf{d}^2.$$

The difference between Covar(FS) and Covar($\bar{P}$O) is therefore

$$(2p^2q^2-2p^2q^2+p^2q^2)\mathbf{d}^2=p^2q^2\mathbf{d}^2=\sigma_\mathbf{D}{}^2/4,$$

so that

$$\text{Covar(FS)}=pq\alpha^2+p^2q^2\mathbf{d}^2=\alpha_\mathbf{A}{}^2/2+\sigma_\mathbf{D}{}^2/4. \qquad (39.4)$$

Dominance variance does enter into the covariance between full sibs. Full sibs can have genotypes as well as genes identical, since each receives one gene from each of the two common parents.

We have repeatedly referred to the influence of identical genes on the genetic covariances between relatives. It should be no surprise, then, that the coefficients of the additive variance component in genetic covariances between relatives are related to the coefficient of coancestry that we discussed in an earlier lecture. It will be recalled that the coefficient of coancestry between parent and offspring was

$$f_{PO}=1/4;$$

between half sibs,

$$f_{HS}=1/8;$$

and between full sibs,

$$f_{FS}=1/4,$$

if the parents were unrelated and noninbred. These coefficients are in each case half of the coefficients of σ_A^2 in the corresponding covariances.

In general, the coefficient of coancestry for a pair of individuals whose parents are neither inbred nor related is half the coefficient of the additive genetic covariance between them. Reversing this statement, $2f_{XY}$ is the coefficient of σ_A^2 in Covar(XY). If X and Y are more distantly related than half sibs, calculating f_{XY} may be simpler than calculating the genetic covariance between them more directly, so that this relationship may be very useful.

If one or more of the parents of X or Y are inbred, $2f_{XY}$ will still be the correct coefficient for σ_A^2 in Covar(XY) (W Nyquist, 1986, personal communication), as long as those parents are unrelated. Inbreeding of, say, X*, a parent of X, will increase f_{X*X}, but will not affect f_{XY} unless X is related to the other parents. If the parents of X and Y are interrelated, however, $2f_{XY}$ will no longer be the correct coefficient.

We have so far ignored σ_I^2. We said earlier that the epistatic variance could be partitioned into a series of subunits:

$$\sigma_I^2=\sigma_{AA}^2+\sigma_{AD}^2+\sigma_{DD}^2+\sigma_{AAA}^2+\ldots$$

In this equation, terms in A only, such as σ_{AA}^2, represent interaction between single gene effects at different loci; terms in D only (σ_{DD}^2), between effects of gene pairs at different loci; and terms in both A and D (σ_{AD}^2), between single gene effects at some loci and gene pair effects at others.

A general rule for the calculation of the coefficients of these components of epistatic interaction variance in the covariance of individuals X and Y can be stated as follows. Let k_A be the coefficient of σ_A^2 and k_D that of σ_D^2 in the genetic covariance between relatives X and Y. A component of the epistatic interaction variance that represents the interactions among single gene effects at x loci and gene pair effects at y loci will have x A's and y D's in its subscript. The coefficient of that component will be $k_A^x k_D^y$. For example, in Covar(PO), σ_A^2 has coefficient 1/2, and σ_D^2, coefficient 0. Therefore, the coefficient of σ_{AA}^2 in Covar(PO) will be $(1/2)^2(0)^0=1/4$. In the same covariance, the coefficient of σ_{AD}^2 will be $(1/2)^1(0)^1=0$, and that of σ_{DD}^2, $(1/2)^0(0)^2=0$. Further examples of the application of the rule are given in Table 39.3.

Thus, the genetic covariance between parent and offspring will contain contributions from effects of interactions between 2, 3, or more loci, in addition to

Table 39.3. Epistatic variance coefficients

	x	y	Covar (PO) $k_A = 1/2$, $k_D = 0$	Covar (FS) $k_A = 1/2$, $k_D = 1/4$
σ_{AA}^2	2	0	1/4	1/4
σ_{AD}^2	1	1	0	1/8
σ_{DD}^2	0	2	0	1/16
σ_{AAA}^2	3	0	1/8	1/8
σ_{AAD}^2	2	1	0	1/16
σ_{ADD}^2	1	2	0	1/32
σ_{DDD}^2	0	3	0	1/64

the sum of the single locus additive variances. It will not contain contributions from dominance variances or from epistatic variance caused by interactions among effects of gene pairs at different loci or among effects of gene pairs at some loci and single genes at others. The genetic covariance among full sibs, on the other hand, will contain contributions from both additive and dominance variances, and from single gene x single gene, single gene x gene pair, and gene pair x gene pair epistatic interaction effects.

The above rule provides some justification for ignoring the contribution of epistatic interaction to covariances between relatives. The coefficients of these components are quite small fractions, and become very small indeed if more than 2 or 3 loci are involved. For example, even if σ_{AAAA}^2 were as large as σ_A^2, the contribution of the four locus interaction to the genetic covariance between parent and offspring would be only 1/8 as great as the contribution of additive variance. The epistatic interaction components tend to be small to begin with; it would be very unlikely that σ_{AAAA}^2 would be nearly as large as σ_A^2 in the first place. When multiplied by the small fractions that are their coefficients, the contributions of these factors to covariance are likely to be considerably less than the sampling error in the estimate.

Also, under long-term selection, linkage disequilibrium tends to cancel epistatic variance (Crow and Kimura 1970). Thus, epistatic interactions between additive effects have little effect on response to selection; σ_A^2, without σ_{AA}^2, σ_{AAA}^2, etc., remains the best predictor of response.

Lecture 39 Exercise

Suppose again that we can identify A_1A_1, A_1A_2 and A_2A_2 through their effect on qualitative traits. In respect to a metric trait, the frequencies of these three genotypes, their values and the value of their offspring are given in the following table:

Genotype	Frequency	P	O
A_1A_1	0.25	1.0	0.6
A_1A_2	0.50	0.2	0
A_2A_2	0.25	−1.4	−0.6

Estimate σ_A^2 from these values.

Lecture 40 Heritability

Bell (1977) traced the use of the term heritability to the 1830's, when it was used in its most literal sense as a qualitative term. A trait which could be inherited bore the quality of heritability.

Johanssen (1903) originated many of our modern genetic terms, including the use of the word "gene" for Mendel's "unit factors". Struck by the difference between the genotype, the genes carried by an individual, and the phenotype, the trait expressed by the individual, he speculated that the phenotype might be determined in part by the genotype and in part by other, nongenotypic effects. He designated the portion of the phenotype determined by the genotype as the heritability for that trait. Lush (1940) recognized and defined two types of heritability: heritability in the broad sense, the ratio of all genetically caused variance to phenotypic variance; and heritability in the narrow sense, the ratio of additive genetic to phenotypic variance. Broad sense heritability is therefore

$$H^2 = \sigma_G{}^2/\sigma_P{}^2. \tag{40.1}$$

Falconer (1981) uses H for broad sense heritability. I have preferred the use of H^2 as analogous to h^2, the universally accepted notation for heritability in the narrow sense.

The concept of broad sense heritability provided a solution to the anthropological debate of the 1920's concerning the relative importance of "nature" and "nurture" in determining human traits, particularly traits of behavior. The proponents of "nature" claimed that such behavior was determined by the innate nature of the human animal; Desmond Morris's *The Naked Ape* (1967) is a late manifestation of the nature side of this controversy. The "nurture" proponents claimed that every newborn child was a "tabula rasa", a blank tablet upon which the rearing of the child could imprint a behavior pattern independent of the child's innate nature. Equating nature with genotype, and nurture with environment, H^2 provides a quantitative answer to this debate. (As with so many disputes between conflicting theories in science, it was found that neither was correct to the exclusion of the other. Both nature and nurture contribute to the establishment of the behavior pattern of any individual.)

Anthropologists and sociologists still retain an interest in the measurement of H^2 (e.g., Mather 1979), using techniques not usually available to the plant or animal breeder, such as comparisons between monozygotic and dizygotic twin pairs.

As we have previously commented, however, plant and animal breeders are usually more interested in the value of a breeding individual's genes than in his own performance, since these will determine the performance of his offspring. Thus, the breeder's interest centers on the additive genetic variance. Wright (1920) seems to have been the first to suggest the use of h^2, the ratio of additive genetic to total phenotypic variance:

$$h^2 = \sigma_A{}^2/\sigma_P{}^2. \qquad (40.2)$$

Lush (1940) termed this ratio "heritability in the narrow sense." He suggested several means of estimating h^2, and indicated the usefulness of this ratio in breeding work. Any reference to heritability in the breeding literature is much more likely to mean h^2 than H^2.

Broad sense heritability is of little use to the animal or plant breeder except that it can provide an upper limit to the value of h^2. Because H^2 includes the sum of additive, dominance, and epistatic variances in its numerator, while h^2 includes only additive, and both have the same denominator, H^2 must be at least as large as h^2. But since in many cases the estimation of H^2 involves at least as much (usually more) effort as estimating h^2, and is less useful, broad sense heritability is seldom estimated by breeders.

One set of circumstances in which the estimation of broad sense heritability may be useful, however, is when a breeder crosses two highly inbred lines, with the objective of selecting from the cross progeny. The parental lines will be homozygous at nearly every locus, and for at least some loci, will carry different alleles. Thus, the parental lines might be

$P_1 = AABBccdd....;$

$P_2 = aaBBCCdd....$

All F_1 offspring will have the same genotype, heterozygous for every locus at which P_1 and P_2 differ:

$F_1 = AaBBCcdd....$

In the F_2 segregation will begin, and a variety of genotypes occur.

Because all F_1 individuals will have the same genotype, there can be no genotypic variation in this generation. All variance must be due to the environment:

$$\mathrm{Var}(F_1) = \sigma_E{}^2.$$

In the F_2 generation, genotypic variance can occur:

$$\mathrm{Var}(F_2) = \sigma_G{}^2 + \sigma_E{}^2 = \sigma_P{}^2.$$

We can therefore estimate broad sense heritability from the phenotypic variances in the two generations.

$$\hat{H}^2 = [\mathrm{Var}(F_2) - \mathrm{Var}(F_1)]/\mathrm{Var}(F_2)$$
$$= (\sigma_G{}^2 + \sigma_E{}^2 - \sigma_E{}^2)/\sigma_P{}^2 = \sigma_G{}^2/\sigma_P{}^2. \qquad (40.3)$$

This may be a useful and relatively inexpensive way of estimating the upper limit for h^2, and therefore of obtaining some idea of the rate of progress that may be obtained in selecting from the hybrid population.

The limitation of this procedure is that it assumes that the environmental variance is the same in both F_1 and F_2. If in fact $\sigma_{E(F1)}{}^2$ and $\sigma_{E(F2)}{}^2$ differ, Eq. (40.3) will not give a valid estimate of the broad sense heritability. There is, in fact, reason to suspect that the two environmental variances may not always be the same.

F_1 individuals are likely to be heterozygous for many loci. In a given F_2 individual, on the average, half the loci heterozygous in its parents will be homozygous. Heterozygous genotypes are more stable under variable environmental conditions than homozygotes, a phenomenon that Lerner (1954) called "genetic homeostasis". [King and Stansfield (1985) have used the term genetic homeostasis to mean the tendency of a population to equilibrate genotypically. Quantitative stability under variable environments they refer to as "developmental homeostasis" or "canalization"]. Even if the breeder controls the environment as closely as he can, there will be microenvironmental differences among individuals of the same generation. These may be sufficient to invoke phenotypic variation among the more homozygous F_2's, but not among heterozygous F_1's. Environmental variance may then be greater in F_2 than in F_1, and Eq. (40.3) will overestimate H^2.

On the other hand, F_1 offspring have homozygous dams from lines P_1 or P_2, while F_2 individuals have heterozygous F_1 dams. Presumably, heterozygous F_1 dams will vary less than the homozygous dams from the parental lines, due to developmental homeostasis. Less variable F_1 dams will provide a less variable maternal environment for their offspring. Maternal environment can have an important effect on the offsprings' phenotypic values, particularly in the case of mammals. As a result, the F_1 offspring may tend to vary more than the F_2. Equation (40.3) may underestimate H^2 if the trait involved is one which is heavily affected by maternal environment.

Frequently, an animal breeder may prefer to initiate a selection program from a cross between existing stocks. The crossbred can be expected to have more genetic variability on which selection can work. Selection cannot be applied before the F_2 generation; F_1 individuals will not be sufficiently varied in genotype to permit successful selection. The breeder might well consider estimating H^2 from F_1 and F_2 variances, as he must produce these generations anyway. The estimate, with all its imperfections, will give him some idea of the h^2 values he may expect in his selection program.

It is more common, of course, to initiate selection from a cross among several inbred strains. To make such a cross, F_1 offspring from pairwise crosses between strains are themselves crossed together. For example, the Purdue University Animal Sciences Mouse Colony maintained an outbred stock which originated from a four-way cross among inbred strains C57BL/6JDt, SWR/DeDt, 129Rr/DgDt, and C3HeB/DeDt (Doolittle and Wilson 1979). The four way cross was made by crossing C57BL/6JDeDt x SWR/DeDt and 129Rr/DgDt x C3HeB/DeDt; F_1 offspring of these two crosses were then crossed to provide the four-way cross offspring.

In such a case, we would be interested in H^2 for the offspring of the cross between the two F_1's; it is this population which will be subjected to selection or other genetic manipulation. But the first generation from the four-way cross will not be genetically uniform; it will segregate at any locus which is heterozygous in either F_1. The phenotypic variance of this generation will include contributions from genotypic variance, and H^2 cannot be estimated by comparing the first with later generations from the four-way cross.

Lecture 40 Exercise

We cross two inbred strains of mice, say, C57BL/6 and C3HeB. F_1 offspring are weighed at 21 days of age; they are later mated at random and an F_2 generation produced. The F_2 mice are also weighed at 21 days old. Variances of 21 day weights are found to be 2.54 g^2 among F_1 and 4.23 g^2 among F_2 individuals. Estimate H^2.

Lecture 41 Realized Heritability

Previously we have described the process of selection on a qualitative trait as if selection were acting directly on the genotype. In fact, the genotype as such is not subject to selection. Differences in fitness associated with different genotypes give rise to selection, but fitness depends on the phenotype associated with a given genotype. The absence of feathers in *nn* male and *n* female carriers of the sex-linked naked gene in fowl (Hutt 1949) causes much of the mortality in brooding pens, the chicks being unable to maintain body temperature without the insulation provided by normal plumage. If extra environmental heat is provided, heat loss is less severe. As a result, mortality, and selection against the naked phenotypes, decreases, without any change in genotype.

Similarly, in individuals homozygous for the recessive gene causing human galactosemia, the enzyme capable of converting galactose to glucose for energy metabolism via the Krebs cycle is lacking. As a result, galactose accumulates in blood and tissues, with serious deleterious effects. If, however, a galactosemic does not eat galactose, there is no accumulation of harmful products. The individual still has the genotype for galactosemia, but avoids the harmful phenotype, the accumulation of galactose; there is then no selection against the galactosemic homozygote.

Quantitative selection is also exerted on phenotypic values. Suppose, for example, that we have a population in which we wish to select for an increase of some metric trait. We will do so by choosing the individuals with the highest phenotypic values in the present generation to serve as parents for the next. The offspring of these individuals should, if the trait is affected by genotype at all, have an average phenotypic value higher than that of the present generation, from which the parents were chosen.

The distribution of phenotypic values in the population in some generation t can be represented by the upper diagram of Fig. 41.1. The average value for this generation is $\bar{y}_t$.

The individuals selected to be the parents of the next generation are represented by the upper tail of the generation t distribution, shaded in Fig. 41.1. We select as many individuals from generation t as needed to make up a predetermined number of matings, starting with that individual who has the largest phenotypic value. In effect, we truncate the distribution at the lowest phenotypic value needed to supply enough mates, using all individuals with that or a higher value. Therefore, this type of selection is sometimes referred to as truncation selection.

The average value for selected individuals is $\bar{y}_{t(s)}$. The selection differential, S, is defined as the amount by which the selected parents exceed the average of the generation from which they were selected. That is, S is the difference between the

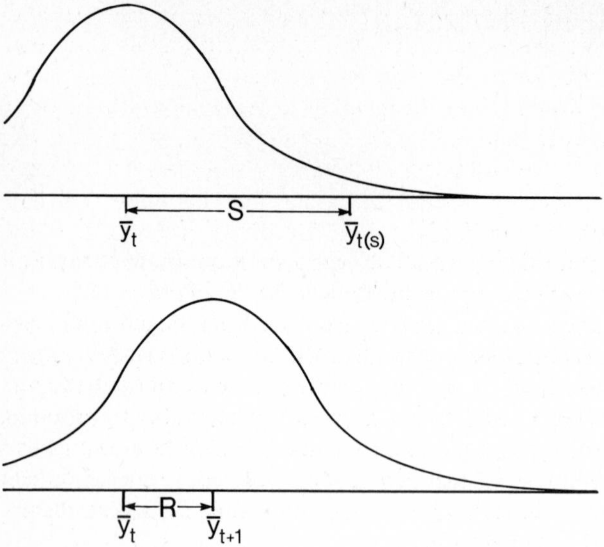

Fig. 41.1. Realized heritability is the ratio of response, the difference between the means of the offspring ($\bar{y}_{t+1}$) and parent ($\bar{y}_t$) populations, to selection differential, the difference between the mean of the selected parents ($\bar{y}_{t(s)}$) and that of the whole parent generation

mean of the selected individuals and the mean of the whole population in generation t:

$$S = \bar{y}_{t(s)} - \bar{y}_t.$$

We make matings among the selected individuals and produce offspring from them to form generation t+1. The distribution of the phenotypic values for generation t+1 is represented by the lower diagram in Fig. 41.1. The mean of this generation is $\bar{y}_{t+1}$. The difference between the means of generations t and t+1 is the response to selection, R:

$$R = \bar{y}_{t+1} - \bar{y}_t.$$

S represents the superiority, in phenotypic values, of the selected parents over the rest of generation t. R represents the superiority, in phenotypic values, of generation t+1 over generation t. But $\bar{y}_{t+1}$, the average of all individuals in generation t+1, is also the average genotypic value of this generation; it therefore represents the average breeding value of the parents selected from generation t. If no selection had been performed, parents for generation t+1 being chosen at random from generation t, $\bar{y}_{t+1}$ would have been equal to $\bar{y}_t$; the average breeding value in generation t would be equal to the average phenotypic value. Thus, R represents the superiority of the selected parents to the rest of generation t in breeding value.

It follows that the ratio R/S is equivalent to the ratio of additive to phenotypic variance; the observed ratio can be used to estimate heritability:

$$\hat{h}^2 = R/S. \qquad (41.1)$$

If the inheritance of the trait were strictly additive, that is, if there were no dominance or epistatic effects on genotypic value, and if environmental variance were negligible, the phenotypic value of an individual would be equal to its breeding value. If S were, say, +8 units, the average phenotypic value of the selected parents would be 8 units greater than the average of generation t as a whole. The average of the breeding values of these selected parents would also be 8 units greater than the average for all of generation t, and R would be +8 as well. The mean of the offspring distribution in Fig. 41.1 would lie directly below that of the selected parents. Then the ratio R/S would be 1. But if all variation in generation t is due to additive genetic variance, $\sigma_P{}^2$ is indeed equal to $\sigma_A{}^2$, and h^2 is 1.

We know, however, that dominance deviations and/or epistatic interactions may, and environmental variance almost certainly will, affect phenotypic values. These effects will tend to decrease the average breeding value of the selected parents, for a given S. Suppose two individuals are identical in genotype except at the A locus, where individual 1 has genotype A_1A_1 and individual 2, A_1A_2. Individual 1 has a greater breeding value than individual 2. But if there were dominance at this locus, the two would have equal genotypic values; if there were positive overdominance, individual 2 would have a larger genotypic value. Similarly, epistatic interactions might give individual 2 a genotypic value as large as or larger than that of individual 1.

Even with strictly additive inheritance at the A locus, so that individual 1 has the higher genotypic value, the phenotypic ranking of the two individuals might be reversed due to different environments. If the environment of individual 1 were unfavorable, and that of individual 2 favorable, the latter might have a larger phenotypic value than the former.

The parents for the next generation are chosen, necessarily, on the basis of their phenotypic values; the distributions in Fig. 41.1 are phenotypic distributions. If, due to nonadditive genotypic or to environmental effects, individual 2 has the same phenotypic value as individual 1, both might be included among the selected parents. If individual 2 has a larger phenotypic value, it might be included among the selected parents, and individual 1 excluded, despite the larger breeding value of the latter. In either case, the average breeding value of the parents would be less than their average phenotypic value. Because it is the average breeding value of the parents that determines the average phenotypic value of their offspring, the center of the offspring distribution will lie to the left of the phenotypic average of the selected parents, as depicted in Fig. 41.1. Then R will be less than S, and h^2 less than 1.

Estimates of heritability obtained from Eq. (41.1) are referred to as realized estimates; they are realized in the sense that they are made real by selection. For purposes of predicting the results of selection, realized heritability estimates are the most useful of estimates. This, of course, amounts to saying that if we want to know what will happen when we select for a given trait, the best way to find out is to select.

In most selection studies, selection will be continued beyond a single generation. We will select the upper tail of the generation t+1 distribution, mate these individuals together to produce generation t+2, select the upper tail of that distribution, etc. From each generation we can obtain an S value, from each successive pair of generations, a value for R; we can then obtain a series of estimates of h^2. The R and S values, and the estimates of heritability, are likely to vary considerably from generation to generation, and we obviously need some way to average these values and/or estimates to produce an estimate based on all the data. Hill (1972a, b) considered several procedures that have been proposed to produce such an average value, and showed that the regression of cumulative response on cumulative selection differential gave the best results under most circumstances.

In our consideration of selection at a single locus, we learned that selection tends to increase the frequency of one allele at the expense of the other. Eventually, the favored allele will become fixed, the other will be eliminated. Selection for a quantitative trait has a similar effect on allele frequencies. Suppose that we are selecting for increased body size. The frequencies of plus alleles at all loci will increase, eventually becoming fixed.

It would seem that selection should therefore lead to a decreasing amount of additive genetic variance, and a decreasing heritability, as the number of loci segregating for alleles with different effects on body size is exhausted. Yet we have above advocated the averaging of h^2 estimates over a series of generations, implying that the heritability parameter is nearly constant.

Heritability does indeed eventually decrease with selection, reaching 0 when all loci are fixed for plus alleles. But the decrease does not begin before many generations of selection have passed. Heritability remains constant for a long period before the decline begins.

In the first place, since many loci are involved, the selection coefficient for any one locus is small. Thus, changes in allele frequencies at any one locus in a single generation are slight. The second reason for the relative constancy of the heritability parameter during the early part of the selection period lies in the nature of the effect of allele frequencies on additive genetic variance.

A given locus will have its maximum effect on additive genetic variance when plus and minus alleles are each at frequency 0.5. Under selection for plus alleles, the effect of loci where $p_t < 0.5$ will be to increase additive variance from generation t to generation $t+1$, as p approaches 0.5. The effect of loci where $p > 0.5$ will be to decrease additive variance from t to $t+1$, as p moves away from 0.5. These increases and decreases tend to cancel, so that additive variance remains more or less constant until p has been raised to 0.5 or more at all loci affecting the trait.

Lecture 41 Exercises

1. In selection for a certain metric trait, suppose that we observe the following mean phenotypic values in a series of generations:

t	$\bar{y}_t$	$\bar{y}_{t(s)}$
0	1.98	2.82
1	2.40	4.98
2	4.69	7.25
3	6.22	9.40
4	7.56	

Calculate estimates of h^2 from each generation of the selection process and an overall average estimate from the regression of responses on selection differentials. (Data adapted from Doolittle 1979).

2. Show that the additive genetic variance at the A-locus is at a maximum when $p=q=0.5$.

Lecture 42 Other Estimates of h^2

Frequently, we may wish to estimate h^2 without doing a selection experiment, or before we begin to select. Under these circumstances, we can use the covariance between relatives to estimate h^2. As we have seen, the covariance of parent with offspring, or of midparent with offspring, or that among half sibs, is a function of $\sigma_A{}^2$. The covariance among full sibs represents a function of both additive and dominance variance components. If we know from other sources that dominance variance is negligible, the full sib covariance can provide another estimate of additive variance.

Estimates of the covariances between parent and offspring or midparent and offspring are simple functions of the measured phenotypic values of these related individuals. The estimate of the covariance of sire and offspring, for example, is

$$\mathrm{Covar(SO)} = [\Sigma(S_iO_i) - (\Sigma S_i)(\Sigma O_i)/n]/(n-1) = \mathrm{SCP(SO)}/(n-1), \qquad (42.1)$$

where S_i is the phenotypic value of the i-th sire, O_i that of his offspring, and there are n sire offspring pairs. Covar(DO), the dam offspring covariance, and Covar($\bar{P}$O), the covariance of the midparent and offspring values, are estimated in a similar manner.

In Eq. (42.1), SCP stands for the sum of the crossproducts, defined as shown in the equation. The abbreviation SSQ will be used to represent the sum of squares; substituting S_i for O_i in Eq. (42.1) would give SSQ(S):

$$\mathrm{SSQ(S)} = \Sigma(S_i{}^2) - (\Sigma S_i)^2/n.$$

We have previously mentioned that the mean of the offspring is closer to the mean of the parental generation than is the mean of the selected parents;

$$R = \bar{y}_{t+1} - \bar{y}_t < \bar{y}_{t(s)} - \bar{y}_t = S$$

in absolute value, unless the phenotypic value is completely determined by the additive genetic value. The offspring generation "regresses" toward the mean of the parent population, relative to the mean of the selected parents. The coefficient of the regression of offspring on parent values therefore provides an estimate of heritability.

The estimate of the regression coefficient for offspring on sire values, for example, is

$$b_{OS} = \mathrm{SCP(SO)/SSQ(S)} = \mathrm{Covar(SO)/Var(S)}.$$

But

$$\mathrm{Covar(SO)} = \hat{\sigma}_A{}^2/2,$$

and because the sire values are phenotypic values, the variance among sire values is

$$Var(S) = SSQ(S)/(n-1) = \hat{\sigma}_P^2.$$

We assume the same σ_P^2 in both sires and offspring. Therefore, the regression is equal to one half of the heritability,

$$b_{OS} = (\hat{\sigma}_A^2/2)/\hat{\sigma}_P^2 = h^2/2,$$

and we can estimate

$$h_{OS}^2 = 2b_{OS}. \tag{42.2a}$$

Similarly,

$$h_{OD}^2 = 2b_{OD}. \tag{42.2b}$$

The variance of the midparent value is half of the phenotypic variance, since each midparent value is the average of two observations. Therefore, the regression of offspring on midparent values is

$$b_{O\bar{P}} = (\hat{\sigma}_A^2/2)/(\hat{\sigma}_P^2/2) = h_{O\bar{P}}^2. \tag{42.2c}$$

Data to estimate regressions of offspring on parental or midparental values can be sampled from a population if parents are chosen and paired at random. Frequently, it will be desirable to expand a population before initiating selection, using one or more generations of random mating; preliminary estimates of heritability can be obtained during this expansion phase. Selection or nonrandom mating may alter variances and covariances in such a way that estimates of heritability from parent-offspring covariances are not valid. For example, the variance of selected parents is not a valid estimate of σ_P^2. Caution must also be used if the phenotypic variances in the two sexes differ. In such a case, the phenotypic variance among sires may be a valid estimate of the phenotypic variance among male, but not among female, offspring. It may be necessary to regress male offspring on sires, female offspring on dams. In some cases, it may be possible to combine sexes if, say, female values are adjusted to male equivalents.

In an animal species where single offspring are the rule, as in cattle or horses, parent-offspring regression procedures may give the best estimates of h^2 available. Parent-offspring regressions can also be estimated from data obtained from polytocous (litter-bearing) mammals, such as pigs and mice. The value of a single offspring can then be replaced by the mean value of a litter of offspring. This mean is a better estimate of the average genotypic value of the offspring than is a single offspring value, because

$$\sigma_{\bar{X}}^2 = \sigma_X^2/n.$$

By the same token, we cannot assume the same phenotypic variance among individual parental values as among litter means. We could and did assume that single parental values and single offspring values had the same variance, in deriving Eqs. (42.2). These estimators must be adjusted, in the case of litter data, for the fact that variances are based on different numbers of values in the two generations. With litter data, it may be better to apply a hierarchical or nested analysis of variance to offspring values, obtaining heritability estimates from half sib and full sib intraclass correlations.

The intraclass correlation coefficient, t, expresses the between group variance as a fraction of the total variance:

$$t = \sigma_B^2/\sigma_T^2.$$

Because the total variance is the sum of the variances between and within groups,

$$t = \sigma_B^2/(\sigma_B^2 + \sigma_W^2).$$

As we have already seen, the between group variance estimates the within group covariance. If the groups are half sib families,

$$t_{HS} = \text{Covar(HS)}/\hat{\sigma}_P^2 = (\hat{\sigma}_A^2/4)/\hat{\sigma}_P^2,$$

so that we can estimate heritability as

$$\hat{h}_{HS}^2 = 4t_{HS}. \tag{42.3}$$

The covariance among full sibs also includes the dominance variance:

$$t_{FS} = (\hat{\sigma}_A^2/2 + \hat{\sigma}_D^2/4)/\hat{\sigma}_P^2.$$

The intraclass correlation among full sibs can therefore be used to estimate h^2 only if the dominance variance is negligibly small. The estimate from this correlation, $2t_{FS}$, can, of course, be compared to other estimates of h^2 to determine whether the dominance variance is significant.

The analysis of variance can be used to estimate components of variance caused by factors which are sources of variation in the analysis, if it is assumed that the levels of the factors are randomly chosen. For example, if we choose and pair sires and dams at random, components of variance due to sire and to dam effects on metric traits of the offspring can be estimated (e.g., Lerner 1950, King and Henderson 1954). These components estimate the variance between sire or dam families, which is, of course, equivalent to the covariance within such families.

Suppose that we have s males to use as sires. Each sire is mated to d dams, no dam being mated to more than one sire, and we produce k offspring from each dam. The hierarchical analysis of variance provides mean square estimates of the variances among the means of sire families; among the means of dam families within each sire; and within each dam family. The mean square among sires is a linear function of three component variances: the variance among sire families, Var(SF); the variance among dam families within sire families, Var(DF); and the variance within dam families, Var(WF). The between dams mean square is a linear function of the latter two components; the within family mean square directly estimates Var(WF). These functions are shown in Table 42.1, which combines and restates Tables 10.3 and 10.4 from Falconer (1981). Var(SF), Var(DF), and Var(WF) can therefore be estimated from functions of the mean squares, as shown in Table 42.1.

All members of the same dam family are full sibs. Members of different dam families within the same sire family are half sibs. Therefore, the variance among sire families estimates the half sib covariance; the variance among dam families

Table 42.1. Hierarchical analysis of variance for half and full sib data

Analysis of variance Source	df	Ex(MSq)
Between sires	s−1	Var (WF)+k Var (DF)+dk Var (SF)
Between dams within sires	s(d−1)	Var (WF)+k Var (DF)
Within dam families	sd(k−1)	Var (WF)

Estimation of components:

Var (SF) = [MSq (Sire) − MSq (Dams)]/dk
Var (DF) = [MSq (Dams) − MSq (WF)]/k
Var (WF) = MSq (WF)

Interpretation in terms of causal variances:

Var (SF) = Covar (HS) = $\hat{\sigma}_A^2/4$
Var (DF) = Covar (FS) − Covar (HS) = $\hat{\sigma}_A^2/4 + \hat{\sigma}_D^2/4$
Var (WF) = $\hat{\sigma}_P^2$ − Covar (FS) = $\hat{\sigma}_A^2/2 + 3\hat{\sigma}_D^2/4 + \hat{\sigma}_E^2$
Var (T) = Var (SF) + Var (DF) + Var (WF) = $\hat{\sigma}_A^2 + \hat{\sigma}_D^2 + \hat{\sigma}_E^2$

Heritability estimates:

Sire component: 4 Var (SF)/Var (T)
Dam component: 4 Var (DF)/Var (T)
Combined: 2[Var (SF) + Var (DF)]/Var (T)

within the same sire family estimates full sib covariance. These covariances can be interpreted in terms of causal variance components, as we have seen. Var(WF) is the variance among full sibs; since full sibs are not identical in genotype, it includes functions of the genetic components. It also includes the variance due to environmental factors which differ between full sibs. The sum of Var(SF), Var(DF), and Var(WF), which we can call Var(T) for total, estimates the phenotypic variance;

$$\begin{aligned}\text{Var(T)} &= \text{Var(SF)} + \text{Var(DF)} + \text{Var(WF)}\\ &= \hat{\sigma}_A^2/4 + \hat{\sigma}_A^2/4 + \hat{\sigma}_D^2/4 + \hat{\sigma}_A^2/2 + 3\hat{\sigma}_D^2/4 + \hat{\sigma}_E^2\\ &= \hat{\sigma}_A^2 + \hat{\sigma}_D^2 + \hat{\sigma}_E^2 = \hat{\sigma}_P^2;\end{aligned}$$

we can use Var(T) as the denominator in estimating h^2.

The estimate of heritability from the sire component is 4Var(SF)/Var(T), because Var(SF) estimates $\sigma_A^2/4$. If the estimate from the dam component, 4Var(DF)/Var(T), is vastly greater than the estimate from the sire component, it may indicate a large dominance variance. If the two components give comparable estimates, however, a combined estimate,

$$\hat{h}_C^2 = [2\text{Var(SF)} + 2\text{Var(DF)}]/\text{Var(T)},$$

may be obtained which will have a somewhat smaller sampling variance than either component alone.

The above estimation procedure ignores σ_I^2. In essence, we are assuming that epistatic variance is negligibly small. In fact, of course, additive x additive, additive x additive x additive, etc., variances are present in Covar(SF), and these plus epistatic variances due to interactions among dominance effects are present in Covar(DF).

Also, members of the same dam family, having the same dam, may be subject to common environmental effects. This is especially true for full sibs that are also litter mates, from litter-bearing mammalian species. Falconer (1981) notes that the covariance of littermate full sibs, and in fact of full sibs in general, may be caused in part by such common environmental factors. We shall have more to say about the effects of common environmental factors in a later lecture.

The hierarchical analysis of variance can also be used in species where one insemination yields only one offspring, if the matings are maintained so that full sib families can be obtained. Because of the temporal differences between successive births, however, the environmental variance among full sibs which are not littermates will usually be much larger than that among full sib littermates.

Lecture 42 Exercises

1. Body weights are taken at 42 days of age on a group of mice. From among these 22 randomly chosen pairs are mated and one offspring from each pair is weighed at 42 days old. Variances of the sire, dam, midparent and offspring values, and covariances of offspring with male and female parents and with midparent values, are as follows:

	Sire	Dam	Midparent	Offspring
Variances	2.75	3.53	1.73	5.98
Covariances with Offspring	0.94	1.90	1.42	

Estimate the heritability of 42-day body weights from regressions of offspring on sire, dam and midparent. Assume that female weights have been adjusted to male equivalents, so that there are no sex differences.

2. Body weights at 42 days of age are taken on offspring from 20 sires, each mated to 2 dams, with 2 offspring weighed per dam family. The analysis of variance mean squares are

Sires	12.8630
Dams within Sires	10.3859
Offspring within Dams	3.1546

Estimate the heritability of 42-day body weights from this analysis.

Lecture 43 Environmental Variance

Quantitative geneticists tend to regard environmental variance as "noise" which interferes with their attempts to evaluate genetic parameters. Some understanding of the nature of this noise is necessary, however, if only to avoid or combat its effects.

Any environmental effects common to all members of a population will be incorporated into the population mean, μ. For example, ambient temperature affects tail length in mice, animals grown in higher temperatures having longer tails (Wilson et al. 1972). Two populations maintained at different temperatures will differ in mean tail length. But if the effect is the same on all members of the same population, temperature will not affect environmental variance within that population.

Only effects which differ from one individual to another within the population contribute to the intrapopulation environmental variance, σ_E^2. Some of these effects will be common to members of a group of individuals, but will differ from one group to another. Others will differ from one individual to another within the same group.

Some effects common to the largest groups may be relatively easy to control, physically or statistically. We can, for example, make sure that all members of a population of animals are maintained under the same temperature. This physical control thrusts the effect of temperature into the population mean by making it common to all members of the population. Or we may repeat the experiment at two or more temperatures, and statistically remove temperature effects in the analysis.

Falconer (1981) distinguishes between two components of environmental variance: a common environmental variance (σ_{Ec}^2) caused by effects common to all members of the same group, and a residual or within environmental variance (σ_{Ew}^2) caused by effects that differ from one individual to another within a group. As the name "residual" for the second of these components implies, total environmental variance is the sum of the two components:

$$\sigma_E^2 = \sigma_{Ec}^2 + \sigma_{Ew}^2.$$

The groups in which the geneticist is interested are usually groups of individuals related in a specific manner, full or half sib families, for example. Any environmental effect common to all members of the same family, but differing from one family to another, will contribute to the variance between families. It will also, therefore, contribute to the covariance within families.

Consider, for example, a litter-bearing mammal such as the mouse. Full sibs who are also littermates occupy the same uterus at the same time. We can therefore expect numerous effects of the uterine environment to be common to members of the same litter. The covariance among full sib littermates will therefore contain a contribution from the common environmental variance;

$$\mathrm{Covar(FS)} = \sigma_A^2/2 + \sigma_D^2/4 + \sigma_{Ec}^2 \tag{43.1}$$

describes this covariance more correctly than Eq. (39.4), which correctly stated the genetic, but ignored the environmental, contribution. Since the environmental contribution to the covariance among full sib littermates arises from their common dam, it is frequently referred to as the maternal effect.

Mammalian full sibs who are not littermates occupy the same uterus, but at different times. Thus, some effects which are common to full sib littermates differ between full sibs who are not littermates, but other effects are common to full sibs, whether littermates or not. We can use Eq. (43.1) to represent the maternal effect for nonlittermate as for littermate full sibs, keeping in mind that σ_{Ec}^2 will probably be less for nonlittermates than for littermates.

Maternal half sib mammals, with the same dam but different sires, also occupy the same uterus at different times, and many of the effects common to nonlittermate full sibs will also be common to these half sibs. A maternal effect contribution to the half sib covariance can be expected, and

$$\mathrm{Covar(HS)} = \sigma_A^2/4 + \sigma_{Ec}^2. \tag{43.2}$$

Equation (43.2) describes the maternal half sib covariance more correctly than Eq. (39.2), again because Eq. (39.2) ignores the common environmental (maternal) effect. Paternal half sibs, with a common sire but different dams, may have few common environmental effects; Eq. (39.2), which is equivalent to Eq. (43.2) with σ_{Ec}^2 set equal to zero, may be an adequate description of paternal half sib covariance.

Mammalian maternal effects are not confined to the common uterine environment. Since the dam feeds and cares for the offspring after birth, postnatal effects can also contribute to the covariance among full or half sibs. Since intrauterine conditions provided by the dam, and her ability to feed and care for her young, are at least partly controlled by her genotype, there is a complex relationship between genetic and nongenetic effects on the covariance among mammalian siblings.

Maternal effects are most obvious, and may be most important, in mammals, especially litter-bearing mammals. But they can exist in other taxa. In birds, for example, the quality of the egg materials which nurture the developing chick is determined by the dam; care and feeding of the nestlings is also at least partly maternal. Thus, the covariances among full and half sibs may again include maternal effects. Even in plants, since the seed constituents are provided by the maternal parent, there may be a maternal effect.

Lecture 44 Repeatability

In many cases, more than one measurement of a given metric trait can be obtained from the same individual. Repetition of measurements on the same individual can be either spatial or temporal. As an example of spatial repetition of measurements, we might measure the length of both the right and left femurs of a mouse, or the size of fruit from two different branches of an apple tree. Temporal repetition might involve body weights taken on successive days on an adult mouse, or leaf area in successive years for a perennial plant.

We must be careful that the organism and trait we have chosen do not show systematic differences between spatially or temporally repeated measures. If we weigh a growing mouse on successive days, its weight on the second day will be systematically larger than on the first day. In effect, the weight of a 22-day-old mouse is not the same trait as the weight of a 21-day-old mouse. In some cases, there is evidence that different genes may act at different ages, e.g., on litter size at different parities in mice. Thus, the body weight of a mouse at 22 days of age might represent the phenotypic expression of a different genotype than that of the same mouse at 21 days of age.

However, let us suppose that such systematic variation does not exist for some trait, so that we can take n measurements of that trait on each of b individuals in a population. The analysis of variance shown in Table 44.1 permits us to divide the total variance among these nb measurements into mean squares among individuals and among measurements within individuals.

We can estimate a within individuals component of variance; this component, σ_W^2, is the expected value of the mean square among measurements within individuals. The among individuals component, σ_B^2, can be estimated by subtracting the among measurements within individuals mean square from the among individuals mean square, and dividing the difference by n:

$$\sigma_B^2 = [\mathrm{MSq(B)} - \mathrm{MSq(W)}]/n. \tag{44.1}$$

Table 44.1. Analysis of variance for multiple measurements on each individual (b individuals, n measurements/individual)

Source of variation	df	Ex(MSq)
Among individuals	$b-1$	$\sigma_W^2 + n\sigma_B^2$
Among measurements within individuals	$b(n-1)$	σ_W^2
Total	$bn-1$	$\sigma_W^2 + [n(b-1)/(bn-1)]\sigma_B^2$

The intraclass correlation among multiple measurements on the same individual has been given a special name; it is called the repeatability, r:

$$r = \sigma_B^2/(\sigma_B^2 + \sigma_W^2). \quad (44.2)$$

Let us look more closely at the nature of repeatability and its components. σ_B^2 measures the variance due to factors that vary from individual to individual, but are common over space and/or time for measurements on the same individual. In the context of our metric inheritance model, if we assume that the members of the population are not all of the same genotype, genotype obviously varies among individuals, but is constant for one individual over space and time. Therefore, σ_G^2 is obviously included in σ_B^2.

There are also, however, environmental effects that, although they vary among individuals, are common to all measurements on the same individual. Falconer (1981) calls the variance due to such effects the general environmental variance, σ_{Eg}^2. Thus,

$$\sigma_B^2 = \sigma_G^2 + \sigma_{Eg}^2. \quad (44.3)$$

Other environmental effects may vary from one measurement to another in the same individual. Falconer calls these special environmental effects. The variance they cause, σ_{Es}^2, is the within individuals variance, σ_W^2.

We can therefore write repeatability as

$$r = (\sigma_G^2 + \sigma_{Eg}^2)/(\sigma_G^2 + \sigma_{Eg}^2 + \sigma_{Es}^2). \quad (44.4)$$

All environmental variance can be classified as either general, varying between individuals but constant for all measurements on the same individual, or special, varying among measurements on the same individual, so that

$$\sigma_{Eg}^2 + \sigma_{Es}^2 = \sigma_E^2,$$

and the denominator of the repeatability ratio is

$$\sigma_G^2 + \sigma_E^2 = \sigma_P^2.$$

It should be noted that the mean square for total variance in the analysis of variance does not estimate the phenotypic variance. Its expected value, from Table 44.1, is

$$\sigma_W^2 + [n(b-1)/nb-1]\sigma_B^2.$$

If b is large, $n(b-1)/(nb-1)$ is very nearly 1, but a slight bias remains if this mean square is used as an estimate of σ_P^2.

The division of σ_E^2 into general and special components is not the same as the division into common and residual components in Lecture 43. The common component of environmental variance reflected effects that differed between individuals from different groups, but were common to all individuals within a group. The residual component measured effects differing between individuals in the same group. The general component we are presently discussing includes all differences between individuals, from the same or different groups. It would thus include both the common and residual components from Lecture 43. The special component is the result of effects which differ from one measurement to another on the same individual.

The repeatability ratio sets an upper limit to the value of H^2 (and therefore also of h^2). Because

$$r=(\sigma_G{}^2+\sigma_{Eg}{}^2)/\sigma_P{}^2 \geqq \sigma_G{}^2/\sigma_P{}^2=H^2,$$

repeatability will be equal to broad sense heritability only if there is no general environmental variance. If general environmental variance exists, repeatability will be larger than H^2. As we have already seen, broad sense heritability is always at least as large as heritability in the narrow sense, so that repeatability must always be as large as or larger than h^2 as well.

An obvious reason for repeating measurements on the same individual is to reduce error variance. The variance of the mean of n observations is 1/n times the variance of a single observation. Thus, the mean of repeated observations on the same individual gives a more accurate measurement of the true value for that individual. While multiple measurements on the same individual do reduce the contribution of $\sigma_{Es}{}^2$ to error variance, however, they do not affect the contribution of $\sigma_{Eg}{}^2$. The variance of the mean of n measurements on the same individual is

$$\mathrm{Var}[\bar{P}(n)]=\sigma_G{}^2+\sigma_{Eg}{}^2+(1/n)\sigma_{Es}{}^2.$$

Repeatability, by permitting us to judge the importance of special environmental variance, allows us to determine whether or not we should repeat measurements on the same individual.

Repeatability will be high if the special environmental variance is small, and low if it is large. Taking repeated measurements on the same individual will reduce error variance very little if repeatability is high; repeated measurements may, however, reduce error variance significantly if repeatability is low.

Falconer (1981) pointed out that dividing the variance of the mean of n observations on a single individual by the phenotypic variance expresses it in terms of repeatability:

$$\begin{aligned}\mathrm{Var}[\bar{P}(n)]/\sigma_P{}^2&=(\sigma_G{}^2+\sigma_{Eg}{}^2)/\sigma_P{}^2+(1/n)(\sigma_{Es}{}^2/\sigma_P{}^2)\\&=r+(1/n)(1-r)=(nr+1-r)/n\\&=[1+r(n-1)]/n.\end{aligned}$$

Thus,

$$\mathrm{Var}[\bar{P}(n)]=\{[1+r(n-1)]/n\}\sigma_P{}^2.$$

As n increases, the effect of $\sigma_{Es}{}^2$ is diluted, and the variance of the mean of observations on the same individual approaches $r\sigma_P{}^2$. $\sigma_G{}^2$ (and $\sigma_A{}^2$) are constant over n, and as n increases they become a larger proportion of $\mathrm{Var}[\bar{P}(n)]$; in effect, the heritability of the mean of n observations on a single individual is greater than that of a single observation, and increases as n increases.

However, $\sigma_{Eg}{}^2$ is also constant over n, and also becomes a greater proportion of the variance of multiple measurements as n increases. Multiple measurements reduce error variance significantly only if $\sigma_{Eg}{}^2$ is a small part of the environmental variance, and $\sigma_{Es}{}^2$ is relatively large. But again, this will only be true if r is small.

In essence, if repeatability is high, we learn nearly as much about an individual from the first measurement on that individual as we can from a series of measurements. Only when r is small do repeated measurements on the same individual give us much additional information.

Even if r is small, the marginal utility, in terms of the reduction of error variance, of taking another measurement on the same individual quickly declines as n increases. Furthermore, increasing the number of measurements on a single individual will usually mean decreasing the number of individuals that can be measured. As a result, there will probably be few occasions when taking more than two or three observations on one individual (on the same trait) will be worthwhile.

Another use of repeatability is to predict future performance. If we have already measured the phenotypic value of an individual with regard to a particular trait, repeatability should help us to predict the result of a second measurement on the same individual. This has no direct genetic connotation, because genotype is constant for one individual over time. The repeatability, r, is an intraclass correlation, and so can be used to estimate the regression of later measurements on earlier ones:

$$\hat{b} = r\sigma_Y/\sigma_X,$$

where X is the first, and Y the second, measurement of the same trait on the same individual. This regression coefficient can then be used to predict future performance of the same individual.

Lecture 44 Exercises

1. Suppose that the phenotypic variance of a single observation per individual for some trait is Var(P) = 100, and that

$$\sigma_A{}^2 = 10 \text{ and } \sigma_E{}^2 = 90,$$

so that

$$\sigma_D{}^2 = \sigma_I{}^2 = 0.$$

The heritabililty for this trait is therefore $h^2 = \sigma_A{}^2/\sigma_P{}^2 = 0.1$. Determine the ratio of $\text{Var}[\bar{P}(n)]/\sigma_P{}^2$, and the heritability of the average when 2, 3, 5 and 10 observations are made per individual, if

a) $\sigma_{Eg}{}^2 = 80$, $\sigma_{Es}{}^2 = 10$, and

b) $\sigma_{Eg}{}^2 = 10$, $\sigma_{Es}{}^2 = 80$.

2. Repeatability of litter size at birth in an outbred stock has been estimated as 0.32. Average litter sizes at 1st, 2nd, and 3rd litters in this stock were 13.25, 15.04, and 14.57 mice, respectively. Predict the number of mice which will be born in second and third litters to a mating whose first litter contains 12 pups.

Appendix I Exercise Solutions

Lecture 1

1. The objective of this exercise is simply to demonstrate that different genotypic distributions in generation t, if they yield the same allele frequencies, will give rise to the same distribution in t+1 and thereafter. Part (a) is used to illustrate the approach; parts (b) through (d) can be solved in a similar manner. The calculation of p_t and q_t for each part of the exercise serves to show that the different populations do indeed yield the same allele frequencies.

a) $D_t: 2H_t: R_t = 0.6: 0: 0.4$

$p_t = D_t + H_t = 0.6 + 0 = 0.6$

Matings	Freq.	Offspring		
		AA	*Aa*	*aa*
AA × *AA*	$D_t^2 = 0.36$	0.36		
AA × *Aa*	$2D_tH_t = 0$			
AA × *aa*	$D_tR_t = 0.24$		0.24	
Aa × *AA*	$2D_tH_t = 0$			
Aa × *Aa*	$4H_t^2 = 0$			
Aa × *aa*	$2H_tR_t = 0$			
aa × *AA*	$D_tR_t = 0.24$		0.24	
aa × *Aa*	$2H_tR_t = 0$			
aa × *aa*	$R_t^2 = 0.16$			0.16
		0.36	0.48	0.16

Hence, $D_{t+1}: 2H_{t+1}: R_{t+1} = 0.36: 0.48: 0.16$.

2. The objective of Exercise 2 is to demonstrate that different allele frequencies in generation t do affect the equilibrium genotypic distribution. Again, only part (a) is done as an example.

a) $D_t: 2H_t: R_t = 0.50: 0: 0.50$

$p_t = D_t + H_t = 0.5$; $q_t = 0.5$

(check: $p_t + q_t = 0.5 + 0.5 = 1$)

$D_{t+1} = p_t^2 = (0.5)^2 = 0.25$;

$2H_{t+1} = 2p_tq_t = 2(0.5)(0.5) = 0.5$;

$R_{t+1} = q_t^2 = 0.25$

(check: $D_{t+1} + 2H_{t+1} + R_{t+1} = 0.25 + 0.50 + 0.25 = 1$)

3. a) Whenever an expression in $xp + yq$ is met, it can be reduced to an expression in p alone, or in q alone, by substituting $q = 1 - p$ or $p = 1 - q$. Thus,

$p + 2q = 1 - q + 2q = 1 + q$; and

$p + 2q = p + 2(1 - p) = p + 2 - 2p = 2 - p$.

b) This example hinges on the fact that

$(p-q)(p+q)=p^2+pq-pq-q^2=p^2-q^2$; therefore,

$p^2-q^2=(p-q)(p+q)=p-q$.

c) Because $p+q=1$, $(p+q)^2=1$; but

$(p+q)^2=p^2+2pq+q^2$, so that

$p^2+q^2=p^2+2pq+q^2-2pq=1-2pq$.

d) Again using the fact that $(p+q)^2=1$,

$(p-q)^2=p^2-2pq+q^2=p^2+2pq+q^2-4pq=1-4pq$.

e) $(p+2q)^2=p^2+4pq+4q^2$; but

$(p+2q)=1+q$, from part a), and therefore

$p^2+4pq+4q^2=(1+q)^2$.

We will encounter these functions again.

Lecture 2

There are, of course, no "correct solutions" for the exercises which accompany Lecture 2. The student may, however, be interested in results obtained by others.

1. To simulate random mating, 10 samples, each of 100 pairs of poker chips, were drawn from bowls containing 5 red, 5 white and 5 blue chips. By replacing each chip drawn, after noting its color and before drawing another chip, samples representative of much larger populations could be obtained using only these few chips. Letting red chips represent *AA*, white *Aa* and blue *aa*, and letting each pair of successive chips represent a mating, results were obtained which are presented in the following table:

	1	2	3	4	5	6	7	8	9	10	Total
AA × *AA*	9	8	8	10	10	14	9	11	11	12	102
AA × *Aa*	10	11	4	5	10	18	10	7	14	12	101
AA × *aa*	13	13	16	12	7	12	17	10	15	5	120
Aa × *AA*	6	13	7	12	11	8	5	17	13	17	109
Aa × *Aa*	13	10	10	13	14	8	12	10	14	13	117
Aa × *aa*	13	17	15	7	13	8	7	13	11	11	115
aa × *AA*	11	10	9	17	10	14	16	13	12	10	122
aa × *Aa*	13	7	18	11	14	9	17	11	5	7	112
aa × *aa*	12	11	13	13	11	9	7	8	5	13	102
											1000

With equal genotypic frequencies, in a sample of 100 pairs we would expect approximately 11 of each of the 9 possible pairs of genotypes. We observed between 4 and 18. Across all ten samples, we would expect each pair of genotypes to occur 111 times; we observed 101 to 122. As expected, the variation was greater in single samples than across all samples, the average absolute deviation being 24% of the expected value in single samples and only 6% across all samples.

If we assume Mendelian segregation from each mating, with each mating producing one offspring, we can calculate the "observed" offspring genotypic ratio. The matings observed would have produced 236 *AA*, 519 *Aa* and 250 *aa*. With D = 2H = R among the parents,

$$p = q = 0.5,$$

and we expect 250 offspring of each homozygous genotype and 500 heterozygotes. There is a slight deficiency in the observed numbers of homozygotes, a slight excess of heterozygotes, but these are not significantly large.

2. To simulate random union of gametes, twelve samples, each of 100 pairs, were drawn from bowls containing red and white chips in the ratio of 5 red(*A*): 10 white (*a*). Thus, $p = 1/3$, $q = 2/3$. Results are presented in the following table:

	1	2	3	4	5	6	7	8	9	10	11	12	Total
$A \times A$	13	15	9	8	8	10	10	14	9	11	11	12	130
$A \times a$	23	21	23	24	20	17	17	30	27	17	29	17	265
$a \times A$	14	20	17	23	16	29	21	22	21	30	25	27	265
$a \times a$	50	44	51	45	56	44	52	34	43	42	35	44	540
													1000

In each sample, 11 *AA*, 22 *Aa*, 22 *aA*, and 44 *aa* were expected; across all samples, 133, 267, 267 and 533. Average absolute deviations represented 14% of expectation in single samples, but only 1% across all samples. There are slight deficiencies in genotypes *AA* and *Aa*, and a slight excess of *aa*, across all samples, but these are not significant.

Lecture 3

1. From these data,

$$N_{Dt} = 342,\ N_{2Ht} = 500,\ N_{Rt} = 187,\ N_{Tt} = 1029;$$

and we can estimate

$$\hat{p}_t = (2N_{Dt} + N_{2Ht})/2N_{Tt} = (684 + 500)/2058 = 0.5753;$$

$$\hat{q}_t = (N_{2Ht} + 2N_{Rt})/2N_{Tt} = 0.4247.$$

The expected numbers of the three genotypes are then

$E_{Dt} = N_{Tt}(\hat{p}_t^2) = 1029(0.5753)^2 = 340.6;$

$E_{2Ht} = 2N_{Tt}(\hat{p}_t)(\hat{q}_t) = 2(1029)(0.5753)(0.4247) = 502.8;$

$E_{Rt} = 1029(0.4247)^2 = 185.6.$

(check: $E_{Dt} + E_{2Ht} + E_{Rt} = 340.6 + 502.8 + 185.6 = 1029 = N_{Tt}$).

We can test the fit of observed to expected numbers:

$$\chi^2 = (342 - 340.6)^2/340.6 + (500 - 502.8)^2/502.8 + (187 - 185.6)^2/185.6 = 0.0319,$$

much less than the critical χ^2 value, 3.841. The null hypothesis is not rejected; the population appears to be in the Hardy-Weinberg ratio of genotypes.

2. In the presence of dominance, allele frequency must be estimated from Eq. (3.5):

$\hat{p}_t' = (342/1029)^{0.5} = 0.5765.$

(Because I^M is the recessive allele in this exercise, and we have been using p_t as the frequency of I^M, we estimate p_t as the recessive allele frequency, although we usually use q_t to symbolize the frequency of the recessive allele.)

The estimated variances of $\hat{p}_t$ and $\hat{p}_t'$ are

$Var(\hat{p}_t) = (0.5753)(0.4247)/2(1029) = 0.000119$, and

$Var(\hat{p}_t') = (-0.3324)/4(1029) = 0.000162$

Thus $\hat{p}_t'$ is more than a third more variable than $\hat{p}_t$.

3. Expected numbers, using the estimate $\hat{p}_t'$ from Exercise 2, are

$E_{Rt+1} = (0.5765)^2(1011) = 336.0;$

$E_{Dt+1} + E_{2Ht+1} = 101 - 336 = 675.0.$

We can test these by χ^2, since in the offspring population we have used only one df, to make the sum of the expected numbers equal to N_t, leaving 1 df for the test.

$$\chi^2 = (340 - 336)^2/336 + (67 - 675)^2/675 = 0.0713.$$

The deviation from expectation is very small, and we conclude that the population is in the Hardy-Weinberg genotypic ratio.

4. We have observed all three genotypes and can therefore, as in Exercise 1, use Eq. (3.1) to estimate allele frequencies:

$N_{Dt} = 457$, $N_{2Ht} = 444$, $N_{Rt} = 231$, $N_{Tt} = 1132$;

$E_{Dt} = 407.2$, $E_{2Ht} = 543.5$, $E_{Rt} = 181.3$;

(check: $407.2 + 543.5 + 181.3 = 1132$);

$$\chi^2 = (49.8)^2/407.2 + (-99.5)^2/543.5 + (49.7)^2/181.3 = 37.9305$$

The null hypothesis is rejected. The population is not in the Hardy-Weinberg ratio of genotypes. There are two few heterozygotes, and too many homozygotes of both types. In fact, the sample was drawn from a strain being inbred by full sib mating while maintaining segregation for c^{ch} and c at the albino locus (Green and Doolittle 1963).

Lecture 4

In this case, we are not given the numbers of parents of each phenotype, and therefore cannot use parental frequencies in estimating q. We are, however, given the total number of recessive offspring, 120, among 843 offspring from all matings. We can therefore use Eq. (3.5) to estimate q:

$$\hat{q}' = (120/843)^{0.5} = 0.3773.$$

From this estimated q, we can estimate Snyder's ratios:

$$\hat{S}_1 = \hat{q}'/(1+\hat{q}') = (0.3773/1.3773) = 0.2739, \text{ and}$$

$$\hat{S}_2 = \hat{S}_1{}^2 = (0.2739)^2 = 0.0750.$$

Applying these ratios, we can derive the expected numbers of black and brown offspring from black x brown and black x black matings, as shown in the table below:

Parents	Offspring				
	Observed		Total	Expected	
	Black	Brown		Black	Brown
Black × black	619	45	664	614	50
Black × brown	104	50	154	112	42
Brown × brown	0	25	25		
Totals	723	120	843		

The observed number of Recessives from brown x brown matings is included in the above table to provide a total number of Recessives from which to estimate q. The numbers of Recessives from black x black and black x brown marriages are independent of the total number of Recessive in the population. Therefore, we can use χ^2 to determine the fit of expected to observed numbers:

$$\chi_1{}^2 = (104-112)^2/112 + (50-42)^2/42 = 2.0952;$$

$$\chi_2{}^2 = (619-614)^2/614 + (45-50)^2/50 = 0.5407.$$

Neither χ^2 is significantly different from zero, and we can conclude that the population is in the Hardy-Weinberg genotypic ratio.

Lecture 5

1. We are given that $p_A = 0.2$, $p_B = 0.1$, and $p_O = 0.7$, respectively. At equilibrium, genotypic frequencies are given by expanding $(\Sigma p_i)^2$; they will therefore be

for L^AL^A: ${p_A}^2 = (0.2)^2 = 0.04$;
L^BL^B: ${p_B}^2 = (0.1)^2 = 0.01$;
L^OL^O: ${p_O}^2 = (0.7)^2 = 0.49$;
L^AL^B: $2p_Ap_B = 2(0.2)(0.1) = 0.04$;
L^AL^O: $2p_Ap_O = 2(0.2)(0.7) = 0.28$;
L^BL^O: $2p_Bp_O = 2(0.1)(0.7) = 0.14$.

Combining these by phenotypic groups:

Group O: 0.49;
Group A: 0.04 + 0.28 = 0.32;
Group B: 0.01 + 0.14 = 0.15;
Group AB: 0.04.

2. In the F_1 of the A x B cross, all mice will be Aa^t heterozygotes. Therefore, the allele frequencies in the three-way cross offspring will be 0.25 A, 0.25 a^t, 0.5 a. Then the genotypes giving an agouti phenotype will be

AA with frequency 0.0625,
Aa^t with frequency 0.125,
Aa with frequency 0.25, and

0.4375 will be the frequency of agoutis.

Genotypes a^ta^t with frequency 0.0625, and
a^ta with frequency 0.25,

0.3125 in all, will have the black-and-tan phenotype. 0.25 of the population will be nonagouti.

Lecture 6

1. We first establish allele frequencies in the two populations. In the male population,

$p_{mt} = 0.4 + 0.2 = 0.6$, $q_{mt} = 0.4$;

in the female population,

$p_{ft} = 0.2 + 0.4 = 0.6$, $q_{ft} = 0.4$.

Both parent populations have the same allele frequencies, although genotypic ratios differ. Thus, the equilibrium ratio will be established in F_1 at

$p^2 : 2pq : q^2 = 0.36 : 0.48 : 0.16$.

The same ratio will also recur in F_2 and F_3.

2. Proceeding in the same manner as in Exercise 1,

$$p_{mt}=0.14;\ q_{mt}=0.86;$$

$$p_{ft}=0.435;\ q_{ft}=0.565.$$

Allelic frequencies are not the same in the two sexes, and the F_1 generation will therefore be in the Bruce ratio:

$$(0.14)(0.435){:}\ (0.14)(0.565)+(0.86)(0.435){:}\ (0.86)(0.565)$$

$$=0.0609{:}\ 0.4532{:}\ 0.4859.$$

Genotypic ratios and allele frequencies will be the same in both sexes of F_1, with

$$p_{F1}=(0.0609+0.2266)=0.2875.$$

Therefore the genotypic ratio in F_2 will be

$$0.0827{:}\ 0.4097{:}\ 0.5076,$$

and this ratio will be repeated in F_3.

3. Some proportion of the matings will be between a male from X and a female from Y, and among F_1 offspring from these matings, the ratio of genotypes will be the same as for the F_1 offspring in Exercise 2. The rest of the matings will be between a male from Y and a female from X; in the offspring of these matings, the Bruce ratio will be

$$(0.435)(0.14){:}\ (0.435)(0.86)+(0.565)(0.14){:}\ (0.565)(0.86)$$

$$=0.0609{:}\ 0.4532{:}\ 0.4859.$$

Thus, whichever of the two reciprocal crosses occurs, the F_1 genotypic ratio will be the same, and identical to that obtained in Exercise 2. As in Exercise 2, the ratio will be

$$0.0827{:}\ 0.4097{:}\ 0.5076$$

in F_2 and F_3.

4. Approximately half of the original matings will be between mates from opposite populations, giving the same offspring ratio as in Exercise 3. One fourth of the matings will be X x X, yielding the genotypic ratio

$$(0.14)^2{:}\ 2(0.14)(0.86){:}\ (0.86)^2=0.0196{:}\ 0.2408{:}\ 0.7396.$$

One fourth will be Y x Y, yielding

$$(0.435)^2{:}\ 2(0.435)(0.565){:}\ (0.565)^2=0.1892{:}\ 0.4916{:}\ 0.3192.$$

The overall ratio in F_1 will therefore be

$$D_{F1}=(0.0196)/4+(0.1892)/4+(0.0609)/2=0.0827;$$

$$2H_{F1}=(0.2408)/4+(0.4916)/4+(0.4532)/2=0.4097;$$

$$R_{F1}=(0.7396)/4+(0.3192)/4+(0.4859)/2=0.5076.$$

The A allele frequency in F_1 is

$$p_{F1} = 0.0827 + 0.2048 = 0.2875,$$

and the genotypic ratio in F_1 is at equilibrium, and will recur in F_2 and F_3.

Lecture 7

The rapid-feathering Leghorn males are all *kk*, the slow-feathering Rhode Island Red females, *K*. Therefore, $p_{mt} = 0$, $p_{ft} = 1$. The average frequency of *K* alleles is, from Eq. (7.5) (with sex roles reversed)

$$\bar{p} = (p_{ft} + 2p_{mt})/3 = 0.3333.$$

But $\bar{p}$ is the equilibrium frequency of *K* in either sex. Therefore, at equilibrium, the genotypic ratio will be

$$D_{fe} = 0.33, \ R_{fe} = 0.67,$$

in females, and in males

$$D_{me} = 0.11, \ 2H_{me} = 0.44, \ R_{me} = 0.45.$$

67% of females and 55% of males will be rapid-feathering.

The successive values of $p_f - p_m$ are:

Generation	$p_f - p_m$	Generation	$p_f - p_m$
0	1	6	0.015625
1	−0.5	7	−0.0078125
2	0.25	8	0.00390625
3	−0.125	9	−0.001953125
4	0.0625	10	0.0009765625
5	−0.03125		

This is less than 5% by generation 5, less than 1% by generation 7, and less than 0.1% by generation 10.

Lecture 8

Sex-linked inheritance of this abnormality appears unlikely, because it passes from father I-1 to son II-1. Holandric inheritance appears to be ruled out because the abnormality passes from grandfather I-1 to grandson III-2 through daughter II-4. The most likely interpretation of the results is that the trait is autosomally inherited, but its expression is either sex-limited or sex-influenced. We let W^a represent the gene for the production of atobase, with an allele W^t for lack of this enzyme. The proband of the pedigree may be either W^tW^t or W^aW^t. II-1 and II-4 are almost certainly both heterozygotes; the lack of the enzyme is expressed in the male but suppressed in the female. This may be due to sex limitation, the enzyme lack being expressed only in males; or sex influence, the W^l gene being dominant in males and recessive in females. III-1 must then be assumed to be a W^aW^a homozygote, while III-2 appears to be another male heterozygote expressing the W^l allele.

Lecture 9

1. BALB/c produces only *Ac* gametes; C57BL/6 only *aC*, so that all F_1 offspring will be *Ac*/*aC*. From Table 9.2, the gametes produced by F_1 will be in the frequencies $g_{11F1} = g_{12F1} = g_{21F1} = g_{22F1} = 1/4$, because the loci are not linked, and will assort independently. From Table 9.1, combining the above gametes at random will give:

	CC	*Cc*	*cc*
AA	1/16	1/8	1/16
Aa	1/8	1/4	1/8
aa	1/16	1/8	1/16

2. The gametic arrays can be calculated from Eqs. (9.2).

Generation	g_{11}	g_{12}	g_{21}	g_{22}	d
a) t	0.13	0.17	0.17	0.53	0.04
t+1	0.11	0.19	0.19	0.51	0.02
t+2	0.10	0.20	0.20	0.50	0.01
t+3	0.095	0.205	0.205	0.495	
b) t	0.13	0.17	0.17	0.53	0.04
t+1	0.12	0.18	0.18	0.52	0.03
t+2	0.1125	0.1875	0.1875	0.5125	0.0225
t+3	0.1069	0.1931	0.1931	0.5069	

Lecture 10

1. In Chapter 9 Exercise 1, allele frequencies in the cross population would be

$g_{11e} = pr = 0.25,$
$g_{12e} = ps = 0.25,$
$g_{21e} = qr = 0.25,$
$g_{22e} = qs = 0.25.$

The array of genotypes at equilibrium would be

	CC	*Cc*	*cc*
AA	1/16	1/8	1/16
Aa	1/8	1/4	1/8
aa	1/16	1/8	1/16

These frequencies are attained in the F_1 generation.

2. Equilibrium gamete frequencies can be calculated from the allele frequencies in generation t:

$$p_t = g_{11t} + g_{12t} = 0.3;$$
$$q_t = g_{21t} + g_{22t} = 0.7;$$
$$r_t = g_{11t} + g_{21t} = 0.3;$$
$$s_t = g_{12t} + g_{22t} = 0.7.$$

Then at equilibrium, $g_{11e} = 0.09$; $g_{12e} = g_{21e} = 0.21$; and $g_{22e} = 0.49$. Therefore, in generation t the disequilibrium is

$$g_{11t} - g_{11e} = 0.04.$$

The same result (ignoring sign) is obtained by taking any difference $g_{ijt} - g_{ije}$; this difference is equal to d_t.

The difference between the frequency of any gamete in generation t+3 and the equilibrium frequency for that gamete is

$$g_{ijt+3} - g_{ije} = 0.005 \text{ if } c = 1/2;$$
$$= 0.0169 \text{ if } c = 1/4.$$

If we calculate d_{t+3} from Eq. (9.2),

$$g_{11t+3}g_{22t+3} - g_{12t+3}g_{21t+3}$$
$$= (0.095)(0.495) - (0.205)^2 = 0.005 \text{ if } c = 1/2;$$
$$= (0.1069)(0.5069) - (0.1931)^2 = 0.0169 \text{ if } c = 1/4.$$

d_{t+3} measures the disequilibrium remaining in generation t+3. d decreases by a fraction c in each generation; therefore, when c is smaller, the decrease per generation is less, and the disequilibrium remaining in generation t+3 is greater.

Lecture 11

In generation 0, all plants are *AAaa*; therefore, from Table 11.1, 2/3 of their gametes will be *Aa*, 1/6 each *AA* and *aa*. This gives the generation 0 gamete frequencies in the table below. d_t represents the gamete change function for generation t, $y_t^2 - x_t z_t$, from Eq. (11.2). The change in frequency for *AA* and *aa* gametes is $+2d_t/3$; for *Aa*, it is $-4d_t/3$. Gamete frequencies for generations 1, 2, and 3 below are calculated from these relationships.

t	x_t	$2y_t$	z_t	d_t
0	0.1667	0.6667	0.1667	0.0833
1	0.2222	0.5556	0.2222	0.0278
2	0.2407	0.5185	0.2407	0.0093
3	0.2469	0.5061	0.2469	

At equilibrium, because $p=q=1/2$, gamete frequencies will be

for *AA* − 0.25;

for *Aa* − 0.50;

for *aa* − 0.25.

The equilibrium genotypic frequencies will then be

AAAA − 0.0625;

AAAa − 0.2500;

AAaa − 0.3750;

Aaaa − 0.2500;

aaaa − 0.0625.

It is apparent that the population rapidly approaches equilibrium. The gamete frequencies in generation 3 are already very near to the equilibrium proportions; and the change function has decreased, by generation 3, to less than 1/3 of 1%.

Lecture 12

1. The zygotic frequency of *mm* in generation t+1 will be q_t^2. Two thirds of these individuals will survive, but only half the survivors will be fertile. These will produce only 80% as many offspring as *MM* and *Mm* individuals. Therefore, the contribution of *mm* to the gametes of generation t+1 will be:

$$(1-s)=(0.6667)(0.5)(0.8)=0.2667.$$

Then the effective genotypic distribution of genotypes in generation t+1 will be [from Eq. (12.1')]:

$$MM\colon Mm\colon mm = p_t^2\colon 2p_tq_t\colon 0.2667q_t^2.$$

From Eq. (12.2'),

$$q_{t+1}=(q_t-sq_t^2)/(1-sq_t^2).$$

But if $1-s=0.2667$, $s=0.7333$, and

$$q_{t+1}=(q_t-0.7333q_t^2)/(1-0.7333q_t^2).$$

2. From Eq. (12.5'),

$$\Delta q_t=-sp_tq_t^2/(1-sq_t^2),$$

and from Eq. (12.4),

$$q_{t+1}=q_t+\Delta q_t.$$

Results for each value of q_t can be arranged in tabular form:

n	q_{t+n}	p_{t+n}	sq_{t+n}^2	Numer.	Denom.	Δq_{t+n}
0	0.25	0.75	0.0458	−0.0344	0.9542	−0.0360
1	0.2139	0.7861	0.0336	−0.0264	0.9664	−0.0273
2	0.1866	0.8134	0.0255	−0.0208	0.9745	−0.0213
3	0.1653					
0	0.4	0.6	0.1173	−0.0704	0.8827	−0.0798
1	0.3202	0.6798	0.0752	−0.0511	0.9248	−0.0552
2	0.2650	0.7350	0.0514	−0.0378	0.9485	−0.0399
3	0.2251					

Lecture 13

In the offspring of generation t, we expect 25 *AA*, 50 *Aa* and 25 *aa* in each sex. We can cull no more than 25 individuals of each sex. If we eliminate all the *aa* individuals in generation t, we cannot cull any *Aa*. That is to say, if we make $s=1$, h must be 0. In order to cull *Aa* individuals, we must leave some *aa* among the breeders (if $h > 0$, s must be less than 1). Because we eliminate 2 *a* alleles with each *aa* individual we cull and only 1 with each *Aa*, it would appear to be the best choice to cull all of the *aa* individuals at the expense of saving all of the *Aa* individuals, that is, to let $s=1$ and $h=0$.

For $s=1$, $h=0$, the change in q is, from Eq. (13.7a),

$$\Delta q_t = -q_t^2/(1+q_t) = -0.25/(1+0.5) = -0.1667.$$

Any other s, h pair will give a smaller negative Δq_t. If we culled, for example, 20 *aa* and 5 *Aa* ($s=0.8$, $hs=0.1$, $h=0.125$), from Eq. (13.7c),

$$\begin{aligned}\Delta q_t &= [-sp_tq_t(h-2hq_t+q_t)]/(1-2hsp_tq_t-sq_t^2)\\ &= -(0.2)(0.5)/(1-0.05-0.2) = -0.1333.\end{aligned}$$

Reducing selection against *aa* in order to increase selection against *Aa* decreases the rate of elimination of *a* genes.

In subsequent generations, q will be less; in generation $t+1$, for example, if have we selected in t with $s=1$, $h=0$, q will only be 0.33. We will then expect to produce fewer *aa* homozygotes, and will be able to increase h without either decreasing s or using fewer than 75 matings per generation.

Lecture 14

1. $s_2=1$, because all sickle cell homozygotes die without reproducing. Normal homozygotes produce only 1/3 as many offspring as heterozygotes, so $s_1=0.6667$, $q_e=0.6667/1.6667=0.4$, $p_e=0.6$.

In confirmation,

$$d\bar{W}_t/dq_t = 2s_1 - 2(s_1 + s_2)q_e$$
$$= 2(0.6667) - 2(1.6667)(0.4) = 1.3333 - 1.3333 = 0.$$

2. With malaria eliminated, $s_1 = 0$, and we have selection against the sickle cell homozygote only. s_2 remains equal to 1, and we have

$$\Delta q_t = -s_2 p_t q_t^2/(1 - s_2 q_t^2)$$
$$= -(1)(0.6)(0.4)^2/[1 - (1)(0.4)^2]$$
$$= -0.096/0.84 = -0.1143.$$
$$q_{t+1} = 0.4 - 0.1143 = 0.2857.$$

The frequency of the sickle cell allele will fall rapidly once the disadvantage of the normal homozygote is removed by eliminating malaria.

2. $q_e = 1/(2 - h')$; therefore

h'	q_e
0	0.5
0.1	0.5263
0.2	0.5556
0.3	0.5882
0.4	0.6250
0.5	0.6667
0.6	0.7143
0.7	0.7692
0.8	0.8333
0.9	0.9091
1	1

As h' increases, that is, as the selection against *a* increases, the equilibrium frequency q_e also increases. When $h' = 1$, q_e is also 1. This does not, however, indicate that selecting as strongly against the *aa* homozygote as against the heterozygote will result in elimination of the *A* allele and the superior *AA* homozygote. The condition $h' = 1$ under negative overdominance is equivalent to $h = 1$ with selection for a superior homozygote, and the population will go to fixation for the *A* allele. As we have seen, with negative overdominance the population does not tend to approach q_e.

Lecture 15

1. a) With $p_t = 0.8$, $\mathbf{u} = 5 \times 10^{-6}$, from Eq. (15.3),

$$p_{t+3} = 0.8(1 - 0.000005)^3 = 0.799988;$$
$$p_{t+5} = 0.8(1 - 0.000005)^5 = 0.799980.$$

b) From Eq. (15.4),

$$\mathbf{u} = (\ln p_t - \ln p_{t+n})/n = (2.079455 - 2.079409)/10$$
$$= 0.000046/10 = 4.6 \times 10^{-6}.$$

c) Again from Eq. (15.4),

$$n = (\ln p_t - \ln p_{t+n})/\mathbf{u} = (2.079455 - 2.078833)/(5 \times 10^{6})$$
$$= 622/5 = 124.$$

Obviously, the change in allele frequencies per generation due to mutation will be very small. The mutation rate prescribed for this exercise is in the range usually suggested for major gene mutations. Population geneticists often ignore the effect of mutation, because it is so small.

2. From Eq. (15.8),

$$q_e = \mathbf{u}/(\mathbf{u} + \mathbf{v}) = 0.00005/0.000055 = 10/11 = 0.91.$$

At equilibrium established by forward versus reverse mutation, more than 90% of the genes will be mutants, only 10% normal; which seems to be almost a contradiction in terms.

Lecture 16

1.

	$h=0$ $q_e=(u/s)^{1/2}$	$h=1/2$ $q_e=2u/s$	$h=1$ $q_e=u/s$
$s=1$	0.001	0.000002	0.000001
$s=1/4$	0.002	0.000008	0.000004

Results with $h=1/2$ lie between those with $h=0$ and with $h=1$, but nearer to the latter.

2. We know that $q_e^2 = 7 \times 10^{-5}$, since this is the observed frequency of PKU-afflicted infants. Assuming still that $s=1$, from Eq. (16.3),

$$7 \times 10^{-5} = \mathbf{u}(1 - 7.5 \times 10^{-5})/1$$
$$= 0.00007/0.99993 = 7.0049 \times 10^{-5}.$$

The improvement in accuracy over the result from Eq. (16.4a), the approximate formula, is far less than the confidence limits on our estimates of either the frequency of recessive homozygotes or mutation rates. Our conclusion, that the gene for PKU is held in the population by a balance between selection and mutation, is valid with either calculation.

3. From Eq. (16.2)

$$q_e = \mathbf{u}(1 - 2hsp_eq_e - sq_e^2)/s(h - 2hq_e + q_e)$$
$$= \mathbf{u}(1 - 2hsp_eq_e + 2hsq_e^2 - sq_e^2)/s(h - 2hsq_e + q_e)$$
$$= \mathbf{u}(1 - 0.25q_e)/(0.25)(0.5);$$
$$(0.25)(0.5)q_e + 0.25\mathbf{u}q_e = \mathbf{u};$$
$$q_e = \mathbf{u}/(0.25)(0.5 + \mathbf{u}).$$

If $\mathbf{u} = 10^{-6}$, then

$$q_e = 10^{-6}/(0.25)(0.500001) = 7.99984 \times 10^{-6}.$$

This result agrees to five decimal places with the result in Exercise 1.

Lecture 17

1. For each population, we must first calculate q^*_t, the average frequency among the other four populations. Then

$$q_{t+1} = q_t - m(q_t - q^*_t),$$

with $m = 0.1$ and q_t as given in the exercise.

Population	q_t	q^*_t	$m(q_t - q^*_t)$	q_{t+1}
A	0.1	0.6	−0.05	0.15
B	0.3	0.55	−0.025	0.325
C	0.5	0.5	0	0.5
D	0.7	0.45	+0.025	0.625
E	0.9	0.4	+0.05	0.85

q_e will be the same for all populations, equal to the average *a* frequency over all populations in generation t, 0.5.

2. $m = 2/3$, so that

$$q_{A,t+1} = q_{At} - 2[q_{At} - (q_{Bt} + q_{Ct})/2]/3$$
$$= q_{At} - 2[2q_{At} - q_{Bt} - q_{Ct}]/6$$
$$= (6q_{At} - 4q_{At} + 2q_{Bt} + 2q_{Ct})/6$$
$$= (q_{At} + q_{Bt} + q_{Ct})/3.$$

The reproducing population in A will consist of 1/3 members of A, 1/3 members of B and 1/3 members of C, and the same will be true of B and of C. In all three populations, allele frequencies will be the average of the initial allele frequencies. The populations will attain equilibrium in generation t+1. This is another case of mixing populations before reproduction takes place; three populations, in this case, are mixed in equal proportions.

Lecture 19

1. The new inbreeding is 1/2N, the old retained, $[1-(1/2N)]F_{t-1}$; F_t is the sum of these.

	N=2			N=10			N=100		
	New	Old	F_t	New	Old	F_t	New	Old	F_t
1	0.25	0	0.25	0.05	0	0.05	0.005	0	0.005
2	0.25	0.1875	0.4375	0.05	0.0475	0.0975	0.005	0.0050	0.0100
3	0.25	0.3281	0.5781	0.05	0.0926	0.1426	0.005	0.0099	0.0149
4	0.25	0.4336	0.6836	0.05	0.1355	0.1855	0.005	0.0148	0.0198
5	0.25	0.5127	0.7627	0.05	0.1762	0.2262	0.005	0.0197	0.0247

F in the line with N=2 increases most rapidly, because of the large amount of new inbreeding. But the proportions of the old inbreeding retained is much less than in the larger lines. Almost all the old inbreeding is retained when N=100.

2.

t	New	Old	F_t
1	0.25	0	0.25
2	0.25	0.1875	0.4375
3	0.25	0.3281	0.5781
4	0.005	0.5752	0.5802
5	0.005	0.5773	0.5823

F_t reaches 0.58 in three generations with N = 2; it is still essentially the same after 2 additional generations with N = 100. Most of the old inbreeding is retained, after the expansion of the lines, but new inbreeding accumulates very slowly.

3. By Eq. (19.2), with $F_0=0$,

$$F_t=1-[1-(1/2N)]^t$$

t	N=2	N=10	N=100
5	0.7627	0.2262	0.0248
10	0.9437	0.4013	0.0489
20	0.9968	0.6415	0.0953

The purpose of this exercise is to demonstrate that F values can be calculated over a large number of generations without calculating all intervening generations, through the use of Eq. (19.2). Values obtained for generation 5 indeed agree with the results from Exercise 1.

Lecture 20

In the results below, the formula is given, followed by the results for $N=2$. A second line gives the results for $N=10$.

1. $\Delta F = 1/2N = 0.25$
$= 0.05$

$1-(1/2N) = 0.75$
$= 0.95$

$[1-(1/2N)]^{10} = (0.75)^{10} = 0.0563$
$= (0.95)^{10} = 0.5987$

$F_{10} = 1-[1-(1/2N)^{10} = 0.9437$
$= 0.4013$

2. $D_{10} = p_0{}^2 + p_0 q_0 F_{10} = (0.6)^2 + (0.6)\,(0.4)\,0.9437) = 0.5865$
$= (0.6)^2 + (0.6)\,(0.4)\,(0.4013) = 0.4563$

$2H_{10} = 2p_0 q_0 (1-F_{10}) = (0.48)\,(0.0563) = 0.0270$
$= (0.48)\,(0.5987) = 0.2874$

$R_{10} = q_0{}^2 + p_0 q_0 F_{10} = (0.4)^2 + (0.6)\,(0.4)\,(0.9437) = 0.3865$
$= (0.4)^2 + (0.6)\,(0.4)\,(0.4013) = 0.2563$

Less than a tenth as much heterozygosity remains in generation 10 when $N=2$ than when $N=10$. Similarly, the frequencies of homozygotes of both types are much closer to their final values when $N=2$.

Lecture 21

1. From Eq. (21.1), $\bar{q}_{it} = q_0$ for any t; therefore,

$\bar{q}_{i1} = \bar{q}_{i2} = \ldots = \bar{q}_{i5} = 0.4.$

From Eq. (21.2),

$V_t(q_i) = p_0 q_0 / 2N + (1 - 1/2N) V_{t-1}(q_i).$

The first term in this equation will be the same in every generation:

$p_0 q_0 / 2N = (0.6)(0.4)/2(5) = 0.024,$

and

$1 - 1/2N = 0.9.$

Therefore,

$V_1(q_i) = 0.024;$

$V_2(q_i) = 0.024 + (0.9)(0.024) = 0.0456;$

$V_3(q_i) = 0.024 + (0.9)(0.0456) = 0.0650;$

$V_4(q_i) = 0.024 + 0.0585 = 0.0825;$

$V_5(q_i) = 0.024 + 0.07425 = 0.0983;$

From Eq. (21.3),

$V_5(q_i) = p_0 q_0 (1 - X^5).$

$X^5 = (0.9)^5 = 0.5904;$

$1 - X^5 = 0.4096;$

$V_5(q_i) = (0.6)(0.4)(0.4096) = 0.0983.$

2. From Eq. (21.3) raised to the limit,

$V_L(q_i) = p_0 q_0 = 0.24.$

Lecture 22

1. We have already calculated $V_5(q_i) = 0.0983$; from Eq. (21.3),

$V_{20}(q_i) = p_0 q_0 (1 - X^{20}) = (0{,}6)\,(0.4)\,(0.8784) = 0.2108.$

Then, from Eq. (22.6), the frequencies of the three genotypes should be, in generation 5:

$p_0{}^2 + V_5(q_i) = 0.36 + 0.0983 = 0.4583,$
$2p_0 q_0 - 2V_5(q_i) = 0.48 - 0.1966 = 0.2834,$
$q_0{}^2 + V_5(q_i) = 0.16 + 0.0983 = 0.2583,$

and in generation 20:

$0.36 + 0{,}2108 = 0.5708,$
$0.48 - 0.4216 = 0.0584,$
$0.16 + 0.2108 = 0.3708.$

2. From Eq. (22.7), in generation 20, the porportions of lines

fixed for *A* will be $0.6 - 3(0.6)\,(0.4)\,(0.1216) = 0.5124,$
fixed for *a* $0.4 - 3(0.6)\,(0.4)\,(0.1216) = 0.3124$, and
segregating $6(0.24)\,(0.1216) = 0.1751.$

Lecture 23

1. From Eq. (23.2),

$N_e = [(1/4)\,(1/N_m + 1/N_f)]^{-1} = [(1/4)\,(1/5 + 1/25)]^{-1}$
$= 1/(0.25)\,(0.20 + 0.04) = 16.67;$

Therefore,

$\Delta F = 1/(2)\,(16.67) = 0.03000.$

2. (a) If one female dies,

$N_e = 1/(0.25)(0.20 + 0.042) = 16.53;$
$\Delta F = 1/(2)(16.53) = 0.03025.$

The increase in F from the loss of one hen is less than 1% of the original inbreeding rate.

(b) If one male dies, however,

$N_e = 1/(0.25)(0.25 + 0.04) = 13.79;$
$\Delta F = 0.03625.$

The increase is nearly 21% of the original inbreeding rate.

3.

	Mass	W/n family
	0	0
	0	2
	1	2
	4	2
	5	4
$\Sigma k_i = N$	10	10
$\bar{k}$	2	2
σ_k^2	5.5	2
N_e	5.1	9.5
ΔF	0.0987	0.0526

The data above is arranged as in Table 23.1. Within family selection reduces the inbreeding rate by nearly half compared to mass selection.

Lecture 24

From Eq. (24.4), $N = 1/(2)(\mathbf{u} + \mathbf{v})$ would prevent significant fixation. Therefore,

$N = 1/(2)(0.011) = 45.5;$

a line with effective breeding size of 45.5 would not be expected to suffer significant fixation at these mutation rates.

Lecture 25

From Eq. (25.4),

$F_C = f_{AB}, F_L = f_{JK}.$

Since in this case, $f_{AB} = f_{JK} = 0$,

$F_C = F_L = 0.$

From the second averaging rule,

$$f_{CL}=(1/4)(f_{AJ}+f_{BK}+f_{AK}+f_{BL})=0.$$

Lecture 26

1. Since A and J are full sibs, from Eq. (26.5),

$$f_{AJ}=(1/4)[1+(F_A/2)+(F_J/2)+f_{AJ}]=1/4.$$

Then, from Eq. (25.6), with f_{BK}, f_{AK} and f_{BJ} all equal to 0,

$$f_{CL}=(1/4)(f_{AJ}+f_{BK}+f_{AK}+f_{BJ})$$
$$=(1/4)(1/4+0+0+0)=1/16.$$

2. As above, $f_{AJ}=1/4$; but now $f_{BK}=1/4$ as well, so that

$$f_{CL}=(1/4)(1/4+1/4+0+0)=1/8.$$

Exercise 1 gives the coancestry of first cousins, Exercise 2 that of double first cousins.

Lecture 27

1.

	A*	J*	B*	K*	A* × J* A	A* × J* J	B* × K* B	B* × K* K	A × B C	J × K L
A*	0	0	0	0	0.25	0.25	0	0	0.125	0.125
J*	0	0.5	0	0	0.25	0.25	0	0	0.125	0.125
B*	0	0	0.5	0	0	0	0.25	0.25	0.125	0.125
K*	0	0	0	0.5	0	0	0.25	0.25	0.125	0.125
A	0.25	0.25	0	0	0.5	0.25	0	0	0.25	0.125
J	0.25	0.25	0	0	0.25	0.5	0	0	0.125	0.25
B	0	0	0.25	0.25	0	0	0.5	0.25	0.25	0.125
K	0	0	0.25	0.25	0	0	0.25	0.5	0.125	0.25
C	0.125	0.125	0.125	0.125	0.25	0.125	0.25	0.125	0.5	0.125
L	0.125	0.125	0.125	0.125	0.125	0.25	0.125	0.25	0.125	0.5

2.

	A	B	B*	C*	A × B C	D*	A × B D	B × B* E	D* × D F	C* × C G	F × E H	G × H J
A	0.5000	0	0	0	0.2500	0	0.2500	0	0.1250	0.1250	0.0625	0.0938
B	0	0.5000	0	0	0.2500	0	0.2500	0.2500	0.1250	0.1250	0.1875	0.1562
B*	0	0	0.5000	0	0	0	0	0.2500	0	0	0.1250	0.0625
C*	0	0	0	0.5000	0	0	0	0	0	0.2500	0	0.1250
C	0.2500	0.2500	0	0	0.5000	0	0.2500	0.1250	0.1250	0.2500	0.1250	0.1875
D*	0	0	0	0	0	0.5000	0	0	0.2500	0	0.1250	0.0625
D	0.2500	0.2500	0	0	0.2500	0	0.5000	0.1250	0.1250	0.1250	0.1875	0.1562
E	0	0.2500	0.2500	0	0.1250	0	0.1250	0.5000	0.625	0.0625	0.2812	0.1719
F	0.1250	0.1250	0	0	0.1250	0.2500	0.2500	0.0625	0.5000	0.0625	0.2812	0.1719
G	0.1250	0.1250	0	0.2500	0.2500	0	0.1250	0.0625	0.0625	0.5000	0.0625	0.2812
H	0.0625	0.1875	0.1250	0	0.1250	0.1250	0.1875	0.2812	0.2812	0.0625	0.5312	0.2969
I	0.0938	0.1562	0.0625	0.1250	0.1875	0.0625	0.1562	0.1719	0.1719	0.2812	0.2969	0.5312

$F_J = 2f_{JJ} - 1 = 0.0625.$

Lecture 28

Common ancestor	Path	n	m	F_{CA}	Contribution to inbreeding
A	N–L–I–F–C–A–D–G–J–M–N	5	5	0	$(1/2)^9 = 0.0020$
B	N–L–I–G–D–B–E–J–H–M–N	5	5	0	$(1/2)^9 = 0.0020$
B	N–L–I–G–D–B–E–J–M–N	5	4	0	$(1/2)^8 = 0.0039$
C	N–L–I–F–C–G–J–M–N	4	4	0	$(1/2)^7 = 0.0078$
G	N–L–I–G–J–M–N	3	3	1/8	$(1/2)^5(9/8) = 0.0352$
					0.0509

This solution is the same (within rounding error) as that obtained by coancestry.

Lecture 29

Backcrossing has the advantage that only nine generations are required to achieve virtual complete inbreeding, while 20 generations of full sib mating are required to accomplish the same result. This advantage is partly obviated, however, by the problem of identifying carriers. Extra generations of inter se matings must be interpolated between each two generations of backcrossing to identify carriers, while in full sib matings we need only breed in each generation from matings that have produced *hs*/*hs* homozygotes. The interpolation of inter se generations makes the elapsed time for backcrossing 17 generations, not significantly less than the 20 needed in full sib mating.

Backcrossing will place *hs* on a background whose properties are well known. Furthermore, C57BL/6 has been used often as a recipient for mutant genes, so that a comparison between the effects of *hs* and other genes with spotting effects would be facilitated. By full sib mating, we could produce a line segregating for *Hs* and *hs* and homozygous at all other loci; but we might in the process fix other genes with modifying effects on *hs*. Backcrossing would appear to be the better procedure.

Lecture 30

1.

Gener.	*AA*	*Aa*	*aa*
0	0.36	0.48	0.16
1	0.48	0.24	0.28
2	0.54	0.12	0.34
3	0.57	0.06	0.37
Lim.	0.6	0	0.4

2. If $p_t=0.25$, $q_t=0.75$, $D_t=0.0625$, $2H_t=0.375$, $R_t=0.5625$.

Then in generation t,

$f(AA \times aa)=0.0625/0.4375=0.1429;$

$f(Aa \times aa)=0.3750/0.4375=0.8571.$

In generation t+1,

$D_{t+1}=0,\ 2H_{t+1}=0.5714,\ R_{t+1}=0.4286;\ p_{t+1}=0.2857.$

And in generation t+2,

$D_{t+2}=0,\ 2H_{t+2}=0.5,\ R_{t+2}=0.5;\ p_{t+2}=0.25.$

Even though the population starts at the equilibrium allele frequencies, it does not remain there. p actually increases between generations t and t+1, then decreases again for t+2.

Lecture 31

The following table shows the genotypic and phenotypic values associated with the various genotypes under the conditions specified in the exercises.

Genotypes	1. Genotypic values	2. Phenotypic values
$H_1H_1H_2H_2H_3H_3$	+6 mm	+8 to +4 mm
H_3h_3	+6	+8 to +4
h_3h_3	+2	+4 to 0
$H_2h_2H_3H_3$	+6	+8 to +4
H_3h_3	+6	+8 to +4
h_3h_3	+2	+4 to 0
$h_2h_2H_3H_3$	+2	+4 to 0
H_3h_3	+2	+4 to 0
h_3h_3	−2	0 to −4
$H_1h_1H_2H_2H_3H_3$	+6	+8 to +4
H_3h_3	+4	+6 to +2
h_3h_3	+2	+4 to 0
$H_2h_2H_3H_3$	+6	+8 to +4
H_3h_3	+4	+6 to +2
h_3h_3	+2	+4 to 0
$h_2h_2H_3H_3$	+2	+4 to 0
H_3h_3	0	+2 to −2
h_3h_3	−2	0 to −4
$h_1h_1H_2H_2H_3H_3$	+2	+4 to 0
H_3h_3	0	+2 to −2
h_3h_3	−2	0 to −4

Genotypes	1. Genotypic values	2. Phenotypic values
$H_2h_2H_3H_3$	+2	+4 to 0
H_3h_3	0	+2 to −2
h_3h_3	−2	0 to −4
$h_2h_2H_3H_3$	−2	0 to −4
H_3h_3	−4	−2 to −6
h_3h_3	−6	−4 to −8

1. There will then be

6 genotypes with genotypic value	+6 mm,
2	+4
9	+2
3	0
5	−2
1	−4
1	−6

as compared to one +6, three +4, six +2, seven 0, six −2, three −4, and one −6 genotypes if there is no dominance or interaction between loci (see Table 31.2).

2. Seventeen genotypes can have a +4 mm phenotypic value, including environmental variation, under these conditions, compared to only 10 in the absence of dominance and interaction.

Lecture 34

1. $p=0.49+(0.42/2)=0.7$; $q=0.3$.

$\tilde{y}=(1-3)/2=-1$;

$\mathbf{a}=(1+3)/2=2$; $\mathbf{d}=-2+1=-1$;

$\bar{M}=(p-q)\mathbf{a}-2pq\mathbf{d}=(0.7-0.3)(2)+2(0.21)(-1)$

$=0.8-0.42=0.38$.

2. $\alpha=\mathbf{a}+(q-p)\mathbf{d}=2+(-0.4)(-1)=2.4$;

$M(A_1)=0.3(2.4)=0.72$; $M(A_2)=-(0.7)(2.4)=-1.68$.

Lecture 35

Numerical Solutions for Combinations of Parameters

p	a	d	α	A_{11}	A_{12}	A_{22}
0.9	2	1	1.2	0.24	−0.96	− 2.16
		0	2.0	0.40	−1.60	− 3.60
		−1	2.8	0.56	−2.24	− 5.04
	4	1	3.2	0.64	−2.56	− 5.76
		0	4.0	0.80	−3.20	− 7.20
		−1	4.8	0.96	−3.84	− 8.64
	6	1	5.2	1.04	−4.16	− 9.36
		0	6.0	1.20	−4.80	−10.80
		−1	6.8	1.36	−5.44	−12.24
0.5	2	1	2	2	0	− 2
		0	2	2	0	− 2
		−1	2	2	0	− 2
	4	1	4	4	0	− 4
		0	4	4	0	− 4
		−1	4	4	0	− 4
	6	1	6	6	0	− 6
		0	6	6	0	− 6
		−1	6	6	0	− 6
0.1	2	1	2.8	5.04	2.24	− 0.56
		0	2.0	3.60	1.60	− 0.40
		−1	1.2	2.16	0.96	− 0.24
	4	1	4.8	8.64	3.84	− 0.96
		0	4.0	7.20	3.20	− 0.80
		−1	3.2	5.76	2.56	− 0.64
	6	1	6.8	12.24	5.44	− 1.36
		0	6.0	10.80	4.80	− 1.20
		−1	5.2	9.36	4.16	− 1.04

We obtain the results in the table above as follows. Given p, **a** and **d**, we first calculate α:

$$\alpha = \mathbf{a} + (q - p)\mathbf{d}.$$

From we can then calculate the three breeding values,

$$\mathbf{A}_{11} = 2q\alpha,\ \mathbf{A}_{12} = (q - p)\ ,\ \text{and}\ \mathbf{A}_{22} = -2p\alpha.$$

When $\mathbf{d} = 0$, $\alpha = \mathbf{a}$, and is independent of p and q. $\alpha = \mathbf{a}$ again, and is independent of **d**, when $p = q = 0.5$. When p is large, $\mathbf{A}_{11}$ is small, and $\mathbf{A}_{12}$ is negative; when p is small, $\mathbf{A}_{11}$ is large and $\mathbf{A}_{12}$ positive. This is because the $\mathbf{A}_{ii*}$ values are measured as deviations from the population mean. When p is large, $\bar{M}$ will also be large, falling between $\mathbf{A}_{11}$ and $\mathbf{A}_{12}$; when p is small, $\bar{M}$ will lie between $\mathbf{A}_{12}$ and $\mathbf{A}_{22}$. The latter value is always negative, of course, since at least one of the $\mathbf{A}_{ii*}$ must always be less than the average.

Lecture 36

p	a	d	D_{11}	D_{12}	D_{22}
0.9	2	1	−0.02	0.18	−1.62
		0	0	0	0
		−1	0.02	−0.18	1.62
	4	1	−0.02	0.18	−1.62
		0	0	0	0
		−1	0.02	−0.18	1.62
	6	1	−0.02	0.18	−1.62
		0	0	0	0
		−1	0.02	−0.18	1.62
0.5	2	1	−0.5	0.5	−0.5
		0	0	0	0
		−1	0.5	−0.5	0.5
	4	1	−0.5	0.5	−0.5
		0	0	0	0
		−1	0.5	−0.5	0.5
	6	1	−0.5	0.5	−0.5
		0	0	0	0
		−1	0.5	−0.5	0.5
0.1	2	1	−1.62	0.18	−0.02
		0	0	0	0
		−1	1.62	−0.18	0.02
	4	1	−1.62	0.18	−0.02
		0	0	0	0
		−1	1.62	−0.18	0.02
	6	1	−1.62	0.18	−0.02
		0	0	0	0
		−1	1.62	−0.18	0.02

Genotypic values are $\mathbf{a}-\bar{M}$, $\mathbf{d}-\bar{M}$ and $-\mathbf{a}-\bar{M}$; these, minus the breeding values calculated in the Lecture 35 exercise, give the $\mathbf{D}_{ii*}$ values for the present exercise.

Whenever $\mathbf{d}=0$, of course, all $\mathbf{D}_{ii'}$ values are 0. When $\mathbf{d}$ is positive, $\mathbf{D}_{11}$ and $\mathbf{D}_{22}$ are negative and $\mathbf{D}_{12}$ is positive, and vice versa when $\mathbf{d}$ is negative. When $\mathbf{d}$ is positive, the genotypic value of the heterozygote must lie above the regression line; in order that the sum of the dominance deviations be zero, the genotypic values of the homozygotes must then be below that line.

Lecture 37

We have measured a large number of individuals of each genotype. We can therefore assume that most of the environmental effects cancel in the genotypic averages presented in the table. These averages represent reasonable estimates of the genotypic values, G_i.

By Eq. (37.4),

$$G_i = A_i + D_i + I_i;$$

therefore,

$$I_i = G_i - A_i - D_i.$$

But the additive effect of the genotype $U_1U_1V_1V_1$ is

$$A_{11} = \mathbf{A}_{U11} + \mathbf{A}_{V11};$$

and additive effects of the other genotypes can be similarly calculated. D_i is the sum of the **D** values. If then, we subtract from our estimate of G_i the sum across loci of the appropriate $\mathbf{A}+\mathbf{D}$ values, we can estimate the epistatic interaction for each genotype.

For example, for the genotype $U_1U_1V_1V_1$,

$$G_i = 6,$$

$$\mathbf{A}_{U11} + \mathbf{D}_{U11} = 3,$$

$$\mathbf{A}_{V11} + \mathbf{D}_{V11} = 2,$$

and therefore I for that genotype is

$$6 - (3+2) = 1.$$

Calculations for the other genotypes are shown below:

$U_iU_1V_1V_2$	$3-(3+1)=-1$
V_2V_2	$0-(3-3)=0$
$U_1U_2V_1V_1$	$3-(1+2)=0$
V_1V_2	$3-(1+1)=1$
V_2V_2	$-3-(1-3)=-5$
$U_2U_2V_1V_1$	$-3-(-4+2)=-1$
V_1V_2	$-3-(-4+1)=0$
V_2V_2	$-6-(-4-3)=1$

Lecture 38

1. Given α and **d** for each locus, we can calculate the locus by locus additive and dominance variances from Eq. (38.4). The across loci sums are then the required $\sigma_{\mathbf{A}}^2$ and $\sigma_{\mathbf{D}}^2$.

	X-locus	Y-locus	Z-locus	Total
$\sigma_{\mathbf{A}}^2 = 2pq\alpha^2$	0.6912	0.7200	0.1728	1.4400
$\sigma_{\mathbf{D}}^2 = 4p^2q^2\mathbf{d}^2$	0	0.0400	0.1474	0.1874

2. Since the phenotypic standard error is 1.7320, the phenotypic variance is the square of that value, $\sigma_{\mathrm{P}}^2 = 3.0000$. Two thirds of this variance is genetic, $\sigma_{\mathrm{G}}^2 = 2.0000$, so that

$$\sigma_{\mathrm{E}}^2 = 3.0000 - 2.0000 = 1.0000.$$

Also, if the genotypic, additive and dominance variances are 2.0000, 1.5840, and 0.1874, respectively, the epistatic interaction variance must be

$$\sigma_I^2 = 2.0000 - 1.5840 - 0.1874 = 0.2286.$$

Lecture 39

With the given values we can calculate Covar(PO), which is

$$\text{Covar(PO)} = (0.25)(1.0)(0.6) + (0.50)(0.2)(0) + (0.25)(-1.4)(-0.6) = 0.36.$$

Then

$$\sigma_A^2 = 2\text{Covar(PO)} = 2(0.36) = 0.72.$$

Lecture 40

According to Eq. (40.3),

$$\hat{H}^2 = [\text{Var}(F_2) - \text{Var}(F_1)]/\text{Var}(F_2).$$

Therefore,

$$\hat{H}^2 = (4.23 - 2.54)/2.54 = 0.40.$$

Falconer (1960, 1981) quotes 0.35 as an approximate median estimate of h^2 for mouse body weights at 6 weeks of age. The estimate of $H^2 = 0.40$ in this example is not unreasonable, especially because h^2 estimates for body weight tend to increase slightly as the age at weighing decreases. Nonadditive genotypic variance is usually of little importance in mouse body weight. Maternal effects can be significant; in this case, however, they appear to have had little effect.

Lecture 41 Exercise Solutions

1. In each generation, we calculate S and R, then use Eq. (41.2),

$$\hat{h}^2 = R/S,$$

to estimate heritability. In generation 0,

$$S = \bar{y}_{0(s)} - \bar{y}_0 = 2.82 - 1.98 = 0.84;$$

$$R = \bar{y}_1 - \bar{y}_0 = 2.40 - 1.98 = 0.42;$$

and

$$R/S = 0.42/0.84 = 0.50.$$

Proceeding in a similar manner for each generation, we obtain

t	S	R	R/S
0	0.84	0.42	0.50
1	2.58	2.29	0.89
2	2.56	1.53	0.60
3	3.18	1.34	0.42

Obviously, the generation by generation estimates vary considerably. To average them, we estimate the regression:

$$b_{RS} = SCP(RS)/SSq(S).$$

From the data, these estimates are

$$SCP(RS) = 14.4390 - 12.7782 = 1.6608;$$
$$SSq(S) = 24.0280 - 20.9764 = 3.0516;$$
$$b_{RS} = 1.6608/3.0516 = 0.54.$$

2. The breeding values of the three genotypes at the A locus, A_1A_1, A_1A_2 and A_2A_2, are $2q\alpha$, $(q-p)\alpha$, and $-2p\alpha$, respectively. These values will be α, 0 and $-\alpha$, respectively, when $p=q=0.5$. Their mean will be

$$q^2\alpha + 0 - p^2\alpha = 0.25\alpha - 0.25\alpha = 0.$$

Additive genetic variance at this locus is then

$$p^2\alpha^2 + q^2\alpha^2 = 0.5\alpha^2.$$

Breeding values are $2(0.6)\alpha$, 0.2α and $-2(0.4)\alpha$ when $p=0.4$, $q=0.6$. The mean is still 0:

$$(0.16)(1.2)\alpha + (0.48)(0.2)\alpha - (0.36)(0.8)\alpha$$
$$= (0.192 + 0.96 - 0.288)\alpha = 0.$$

The additive genetic variance will then be

$$(0.16)(1.44)\alpha^2 + (0.48)(0.04)\alpha^2 + (0.36)(0.64)\alpha^2$$
$$= 0.48\alpha^2.$$

This will also be the additive genetic variance when $p=0.6$, $q=0.4$, the variance equation being symmetric. Thus, $\sigma_A{}^2$ is at a maximum when $p=q=0.5$, and decreases as either p or q increases.

Lecture 42

1. Because $b_{YX} = Covar(XY)/Var(X)$, we can estimate

$$b_{OS} = 0.94/2.75 = 0.34;$$

$$b_{OD} = 1.90/3.53 = 0.54;$$

$$b_{O\bar{P}} = 1.42/1.73 = 0.82.$$

Therefore, our estimates of heritability are

$h_S^2 = 0.68;$

$h_D^2 = 1.08;$

$h_{\bar{P}} = 0.82.$

where the subscripts S, D, and $\bar{P}$ represent respectively estimates from the regressions on sire, dam, and midparent.

2. Taking differences among the mean squares as indicated in Table 41.1, we obtain estimates of the components:

Var(SF) = 0.6193;

Var(DF) = 3.6156;

Var(WF) = 3.1546.

From these we can estimate heritability:

$\hat{h}_S^2 = 0.3352;$

$\hat{h}_D^2 = 1.9572;$

$\hat{h}_C^2 = 1.1461.$

Heritability represents the fraction of the total phenotypic variance accounted for by additive genetic variance, and therefore cannot be greater than one. The estimate from the analysis of variance dam component is therefore obviously overstated, which could be due to the presence of dominance variance or of maternal effects. As the same effect seems to be present in the estimate from the regression on dam values, it seems more likely that the excessively large estimate is due to the presence of maternal effects.

The sire component estimate from the analysis of variance seems to be within the range (0.3 – 0.5) of most estimates for body weights quoted in the literature. The estimate from the regression of offspring on sires is too large; the estimate of the regression on midparent values would appear to be nearer the normal range. The reason for this is not clear, unless sires were chosen with a bias toward larger mice, or unless there was some extent of nonrandom mating.

Lecture 44

1. $\mathrm{Var}[\bar{P}(n)]/\sigma_P^2 = [1 + r(n-1)]/n.$
Therefore,

a) $r = (10 + 80)/100 = 0.9.$

n	$1 + r(n-1)$	$\mathrm{Var}[\bar{P}(n)]/\sigma_P^2$	$\mathrm{Var}[\bar{P}(n)]$	h^2
2	1.9	0.95	95	0.1053
3	2.8	0.93	93	0.1075
5	4.6	0.92	92	0.1087
10	9.1	0.91	91	0.1099

b) $r=(10+10)/100=0.2$.

2	1.2	0.6	60	0.1667
3	1.4	0.47	47	0.2128
5	1.8	0.36	36	0.2778
10	2.8	0.28	28	0.3571

When r is high, the heritability of the average of n measurements is not much greater than that of a single measurement. In the above example, the increase is only about 10% when 10 measurements are taken. But with low r, a second measurement increases h^2 by 2/3. A third measurement increases h^2 by only about 1/4 over two measurements, and the marginal utility of further measurements decreases rapidly.

2. If X is the first litter size, $\hat{Y}_2$ and $\hat{Y}_3$ the predicted second and third litter sizes,

$$\hat{Y}_i=0.32(X-\bar{X})+\bar{Y}_i$$

$$=0.32(12-13.25)+\bar{Y}_i=-0.4+\bar{Y}_i;$$

$$\bar{Y}_2=15.04,\ \hat{Y}_2=14.64;$$

$$\bar{Y}_3=14.57,\ \hat{Y}_3=14.17.$$

Appendix II Review Lectures

Review Lecture 1 Basic Genetics

Modern genetics began in 1900 with the rediscovery, by de Vries, Correns, and Tschermak, of Mendel's 1865 paper describing the inheritance of the characteristics of peas, by the action of units to which Johanssen later (1903) gave the name "genes". Mendel determined that these units were carried in pairs in the normal individual. During the formation of gametes, the members of a given gene pair separate, each gamete receiving only one member of each pair. When two gametes unite to form a new individual, the pairing is reestablished. Thus, the two genes of any pair in a given individual come to it, one from its sire and one from its dam. The offspring individual will, in turn, pass on one member of its pair to each of its offspring; on the average, half of its offspring will receive its paternal, half its maternal gene.

The genes of any pair can exist in two or more alternate forms, called alleles. Suppose that for a given pair, two alleles exist, which we can designate *A* and *a*. Any individual can then have one of three possible genic constitutions, or genotypes: *AA*, *Aa* or *aa*. An individual which has two like alleles is formed by the union of two gametes that carry alleles that are alike. *AA* is formed by the union of two *A* gametes; *aa*, by the union of two *a* gametes. These two genotypes are said to be "homozygous". The genotype *Aa*, on the other hand, is formed by the union of gametes bearing different alleles, and is said to be "heterozygous".

A homozygous genotype produces only one type of gamete, but the heterozygote produces both types, in approximately equal numbers. Because of these differences in gamete production, homozygotes and heterozygotes produce offspring genotypes in different proportions. A mating between two homozygotes can produce only one genotype in the offspring: $AA \times AA$, only *AA*; $aa \times aa$, only *aa*; and $AA \times aa$, only *Aa*. But a heterozygote mating will always produce at least two offspring genotypes: $Aa \times AA$, equal numbers of *AA* and *Aa*; and $Aa \times aa$, *Aa* and *aa* in equal numbers. When two heterozygotes mate, all three genotypes are produced, in the ratio 1 *AA*: 2 *Aa*: 1 *aa*.

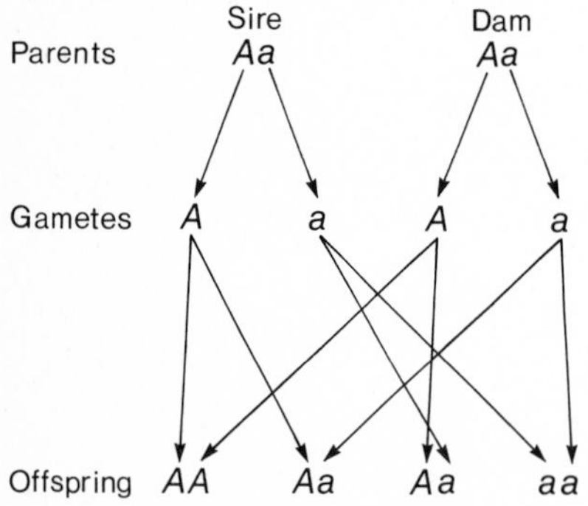

Fig. R1.1. Segregation in the $Aa \times Aa$ mating

The diagram in Fig. R1.1 will help to explain the result of a heterozygote × heterozygote mating. The *A* gamete from one parent can unite with either the *A* or the *a* gamete from the other, producing a *AA* or a *Aa* offspring. The *a* gamete from the first parent can unite with either gamete from the second, producing *Aa* or *aa*. These four gametic union possibilities are equally likely. Thus, the heterozygote × heterozygote mating produces all three genotypes in its offspring. Because two of the possible gametic unions result in the formation of heterozygote offspring, the genotypes are produced in the 1: 2: 1 ratio.

The British pioneer of genetics, R. C. Punnett, invented a tabular form in which to visualize the results of matings of various types, the Punnett square. Table R1.1 shows the results of the mating *Aa* × *Aa* in the form of a Punnett square. Rows represent sire gametes, columns the gametes from the dam; the body of the table then presents the offspring genotypes. It is assumed that each gamete from either parent is equally likely to occur, so each cell in the table has equal probability, 1/4. *Aa* offspring occur twice as frequently as either of the other genotypes because they can occur as the result of two different gametic combinations.

Note that the Punnett square for *Aa* × *Aa* contains within it the squares for the other possible matings with this gene pair. Row 1, column 1 represents the *AA* × *AA* mating; row 2, column 2, the *aa* × *aa* mating. Row 1, column 2 represents the mating between a *AA* sire and a *aa* dam; row 2, column 1 the reciprocal mating. The two columns in row 1 show the results of mating a *AA* sire with a *Aa* dam; the two columns in row 2, those of mating a *aa* sire with a *Aa* dam. Similarly, the two rows in column 1 show the results of mating a *Aa* sire with a *AA* dam, the rows in column 2, a *Aa* sire with a *aa* dam. (The Punnett "square" need not always be square; it will be rectangular if one parent produces more types of gametes than the other.)

Gene pairs like those with which Mendel worked control characteristics or traits of the organism. For example, one of the traits in which Mendel's peas differed from line to line was plant height. He concluded that a pure line of tall plants which never produced dwarfs carried two alleles for tallness; we can symbolize their genotype as *TT*. Similarly, a pure line of dwarf plants carried two of the alternate alleles, *tt*, for dwarfness. When he crossed a tall and a dwarf line, the F_1 offspring were all *Tt* in gene constitution, but were as tall as their *TT* parent. He attributed this to the phenomenon of dominance. If opposite alleles are present in the same individual, sometimes one allele will be expressed, while the other is masked. The allele that is expressed is said to be dominant, the masked allele, re-

Table R1.1. Punnett square for the *Aa* × *Aa* mating

		Dam gametes	
		A	*a*
Sire gametes	*A*	*AA*	*Aa*
	a	*Aa*	*aa*

cessive. The masked allele will, of course, be passed on to half of the offspring of a heterozygote, and will be expressed in any homozygous recessive offspring. The phenotype, the characteristics expressed by the individual, does not always accurately express the genotype, the genic constitution of that individual.

The phenomenon of dominance does not apply to some gene pairs. Sometimes the heterozygote may be phenotypically intermediate between the two homozygotes, for example, pink flowers on the hybrid offspring of a cross between red and white flowered snapdragons (Sinnott et al. 1950). This is referred to as partial dominance, particularly if the heterozygote's phenotype appears closer to that of one homozygote than that of the other. In other cases, the heterozygote may express the characteristics of both homozygotes; this is called codominance. In the human MN blood groups, for example, the homozygote for one allele has the M antigens on his erythrocytes, while the opposite homozygote has the N antigen. The heterozygote has both alleles and produces both antigens (Thompson and Thompson 1980).

There may be more than two alleles for a given gene pair, (although, of course, any one individual cannot carry more than two, at most). The pink-eye dilution series in the house mouse provides an extreme example, with 13 alleles (Silvers 1979). In multiple allele series, there is frequently a dominance hierarchy, with a given allele dominant to those below it, and recessive to those above it, in the hierarchy. In the human ABO blood groups, we have a combination of codominance and dominance; the alleles for the A and B antigens are both dominant to the O allele, but codominant to each other, the AB heterozygote carrying both antigens to constitute the relatively rare AB blood group (Bernstein 1924).

Duplicate genes are different gene pairs with the same or very similar effects on the phenotype. For example, both the blue dilution and the leaden dilution genes in the house mouse cause very similar dilution of pigments in the coat (Silvers 1979). They are, therefore, duplicate genes.

Coat color in the mouse, in fact, as in other species, is a very complex trait, controlled by the action of multiple gene pairs. Silvers's book (1979) of over 300 pages is entirely devoted to the description of these multiple factors, more than 60 gene pairs being discussed.

Mouse coat color also provides an example of the type of interaction between different gene pairs that is called epistasis. Albino mice produce no pigment; therefore, the complex effects of other gene pairs that produce a rich variety of hues and shades are all masked in an albino homozygote. Regardless of the color, location, intensity, and so forth, of pigment that the rest of the coat color genotype may call for, an albino homozygote is uniformly white with pink eyes. The homozygous albino genotype is epistatic to all other coat color genes, which are hypostatic to albino.

Mendel originally proposed, on the basis of his experiments with the garden pea, that gene pairs controlling different traits segregated independently. When he crossed a plant breeding pure for yellow cotyledons with one breeding pure for green cotyledons, the F_2 segregated to give yellow and green cotyledons in a 3:1 ratio, yellow being dominant. Similarly, the F_2 of a cross between a plant with wrinkled and a plant with round seed gave a ratio of 3 wrinkled to 1 round seeded plant. If both traits occurred in the same cross, e.g., if he crossed a plant with yellow cotyledons and wrinkled seed with one with green cotyledons and round seed, the F_2 plants appeared in the ratio 9 yellow wrinkled to 3 yellow round to 3 green wrinkled to 1 green round. He recognized this ratio as indicating that the gene pair for cotyledon color and the pair for seed form were segregating independently.

However, Bateson and Punnett (1906) demonstrated that pairs of traits did not always segregate independently. Sutton (1902) suggested that the genes were associated with the chromosomes; Bridges demonstrated (1914) the association of sex-linked genes with the X chromosome, and later (1916) the association of other gene pairs with the autosomes. Because there are more gene pairs than chromosomes, each chromosome must carry more than one pair; Morgan et al. (1915) suggested a linear arrangement of the genes along the chromosome. Gene pairs on the same chromosome would not be expected to segregate independently.

Genes on the same chromosome are not immutably bound together. Bateson and Punnett (1906) had observed that even when two traits did not segregate independently, they did recombine to some limited extent. Janssens (1909) pointed out that chiasmata, chromosomal crossing configurations observed during meiotic prophase, might represent the exchange of material between sister chromatids, and that such exchanges would permit recombination between gene pairs on the same chromosome. It was not until 1931, however, that Creighton and McClintock demonstrated conclusively the correlation between visually observed chiasmata and genetically demonstrated recombination. With gene pairs ranged linearly, and chiasmata occurring at random points, along the chromosome, it was possible to calculate, from recombination frequencies, the order of, and the distance between, two linked gene pairs (that is, two pairs on the same chromosome). From such data linkage maps showing the locations of gene pairs on each chromosome of a given species (e.g., Green 1966 for the mouse) have been constructed.

The locus (Latin for location) of a gene pair has come to be used as a shorthand designation for the gene pair itself. Thus, we refer to the gene pair controlling black vs. brown hair pigment in the mouse as the "brown pigment locus", etc., although, strictly speaking, the locus is only the location of the gene pair on the chromosomes of the mouse. This shorthand will be used extensively in the text.

The effect of recombination between gene pairs, whether linked or not, on population genetics is discussed in detail in Lectures 9 and 10. Linkage is of little importance in quantitative traits. A large number of gene pairs, scattered throughout the chromosomes, controls the inheritance of such traits, and each pair has only a small quantitative effect.

It is unfortunate that we have not space to detail the brilliant chain of research studies that led to the elucidation of the physico-chemical nature and the metabolic activity of the gene. It must here suffice to say that the gene appears to be a self-reproducing molecule of deoxyribose nucleic acid (DNA); that this molecule consists of a double helical chain of purine and pyrimidine bases, joined together by phosphate and deoxyribose groups; and that the order of the bases in the gene molecule encodes specifications for the structure of metabolic products. Different alleles differ in base order, and a mutation, the process by which one allele changes into another, is accomplished by spontaneous or induced changes in base order.

The chain, or perhaps it is better to say, the network, of research that has elucidated the behavior of genes in populations is no less brilliant. The present text attempts to guide the student through this network. Necessarily, only a few salient points of the results of these studies can be hit upon. The whole story could never be assembled in detail between a single pair of book covers. We stand upon the shoulders of the giants that preceded us, and we hope to provide a solid foundation for those who will follow.

Review Lecture 2 Probability

Whenever an event occurs for which more than one outcome is possible, and chance enters into the determination of the outcome, we can discuss the probabilities of the possible outcomes. The probability of outcome A is the chance that A, rather than B or C or some other possible outcome, will take place. For example, we toss a coin. Possible outcomes are a head or a tail; chance will determine which outcome actually occurs. We can, therefore, discuss the probability that a head (or a tail) will occur.

If the event occurs repeatedly, and we keep a record of the outcomes, we can calculate the frequency of a given outcome. The frequency of outcome A is the proportion of occurrences or trials of the event in which the outcome was A. We can estimate the probability that A will occur in a single trial from the frequency with which A has occurred over a series of trials. Over a series of a few trials, the frequency of A may not give a very good estimate of the single trial probability of A. For example, if our "series" consists of only a single trial, either A did occur on that single trial, or it did not; its frequency is 1 or 0. In any meaningful situation, of course, the probability of A is neither 0 nor 1, but something in between.

As the number of trials increases, however, frequency becomes a better and better estimator of probability. In fact, the probability of outcome A is the limiting value of the frequency of A; if it were possible to record the outcomes of an infinite number of trials, the frequency of A over that series would be exactly equal to the probability of an A outcome in any single trial.

If the event whose probability we wish to evaluate is complex, with many unknown factors affecting the outcome, an estimate of probability deduced a posteriori from empirical observations may be the best, indeed the only, estimate that we can obtain. Often, however, we can induce from the nature of the event an a priori estimate of the probability of a given outcome. In tossing a coin, for example, we know that the coin has two faces. If we believe that the coin is fair, and is not tossed in some manner likely to favor one outcome or the other, we can argue that a head or a tail are equally likely to occur. Then the a priori probability of either outcome is 0.5.

Both probabilities and frequencies are common fractions, with values between 0 and 1. The sum of the number of A outcomes, plus the number of B outcomes, and so forth, must be the total number in the series of trials, because the outcome of any trial must fall into one or another of the outcome classes. Therefore the total of the frequencies of the outcomes must be 1, no matter how many trials are included in the series. This will also be true in the limit, so that the sum of the probabilities of all of the outcomes must also be 1.

Genetic events are stochastic; that is, chance plays a role in determining their outcomes. In general, a priori probabilities for the outcomes of genetic events can be induced from their nature. For example, consider the process of forming offspring from a mating between two *Aa* heterozygotes, illustrated by the Punnett square in Table R1.1. The process involves the choice of one gamete from the sire, one from the dam, and the union of the chosen gametes. The two gamete choices are stochastic events; chance plays a role in determining the results of these choices. Once the gametes are chosen, of course, the result of gametic union is completely determined, and the union *per se* is not a stochastic event.

The heterozygous sire forms two types of gamete, in approximately equal numbers, those carrying *A* and those carrying *a*. If we have no reason to suppose that either type of gamete is more likely to be chosen than the other, the a priori estimate of the probability of choosing a *A* sire gamete is $\Pr(A_s)=0.5$. The a priori probability of choosing a *a* sire gamete is $\Pr(a_s)$, also equal to 0.5. The dam similarly forms approximately equal numbers of *A* and *a* gametes, so that $\Pr(A_d)=\Pr(a_d)=0.5$.

Because the choice of the two gametes that will unite determines the genotype of the offspring, the probability of a *AA* offspring is the probability that a *A* sire gamete and a *A* dam gamete will be chosen at the same time. This probability is the product of the probabilities of the two gamete choices:

$$\Pr(AA)=\Pr(A_s,A_d)=\Pr(A_s)\Pr(A_d)=0.25. \qquad \text{(R2.1a)}$$

Similarly, the probability of a *aa* offspring is

$$\Pr(aa)=\Pr(a_s)\Pr(a_d)=0.25. \qquad \text{(R2.1b)}$$

Finally, there are two ways in which a *Aa* offspring can occur: the union of A_s with a_d, or that of a_s with A_d. Therefore,

$$\begin{aligned}\Pr(Aa)&=\Pr(A_s)\Pr(a_d)+\Pr(a_s)\Pr(A_d)\\&=0.25+0.25=0.5.\end{aligned} \qquad \text{(R2.1c)}$$

The sum of Eqs. (R2.1a, b and c) is

$$0.25+0.25+0.5=1;$$

offspring must be either *AA*, *Aa* or *aa*.

The above argument illustrates the two ways of combining probabilities. To obtain a *AA* homozygote, we must have the simultaneous occurrence of two stochastic events, the choice of a *A* gamete from the sire, *and* the choice of *A* from the dam. The probability that both of these events occur at the same time, the probability of the intersection of these two events, is the product of the probabilities of each of the two events:

$$\Pr(A_s \text{ and } A_d)=\Pr(A_s)\Pr(A_d).$$

To obtain a heterozygous offspring, either of two events can occur, the union of a *A* gamete from the sire with a *a* gamete from the dam, or vice versa. The probability that either of these two events occurs, the probability of the union of these two events, is the sum of the probabilities of each:

$$\Pr(A_s,a_d \text{ or } a_s,A_d)$$
$$= \Pr(A_s, a_d) + \Pr(a_s, A_d).$$

The probability of the intersection of two events is the product, that of their union is the sum, of their separate probabilities (but only if the two events are independent).

Assuming that A is dominant to a, the probability of an offspring of the dominant phenotype is again the probability of the union of two events. A dominant offspring is either one that is AA or one that is Aa. Thus, either outcome will give us a dominant offspring, and

$$\Pr(\text{dominant}) = \Pr(AA) + \Pr(Aa) = 0.25 + 0.5 = 0.75. \qquad \text{(R2.2a)}$$

The probability of a recessive offspring must be

$$\Pr(\text{recessive}) = 1 - \Pr(\text{dominant}) = 1 - 0.75 = 0.25, \qquad \text{(R2.2b)}$$

because any offspring must be either a dominant or a recessive. Only one genotype, aa, will give a recessive offspring, and we have already seen that the probability of this genotype is 0.25 [Eq. (R2.1b)].

If we have a priori probabilities, we can reverse the rule that the large sample frequencies provide an estimate of probability. With the probability established beforehand, we can derive the expected frequency of an outcome. For example, from the probabilities in Eqs. (R2.1) and (R2.2), we can predict the 1: 2: 1 genotypic and 3:1 phenotypic ratios expected among the offspring of two heterozygotes mated together.

In population genetics, an additional level of stochastic events is superimposed upon the stochastic choices of gametes from given parents. Consider, for example, matings among the offspring of the heterozygote x heterozygote mating discussed above. If these matings are made at random any offspring has equal probability of being chosen as sire or dam for any given mating. Thus, random mating among the offspring of a Aa x Aa mating means that the probabilities of choosing a AA, a Aa or a aa sire are 0.25, 0.5 and 0.25, respectively.

The choices of the sire and of the dam are independent of one another under random mating. Therefore, the probabilities of choosing a AA, a Aa or a aa dam are also 0.25, 0.5 and 0.25, regardless of the genotype of the sire chosen. The probability of a given type of mating is then the product of the probability of choosing the appropriate sire genotype, times the probability of choosing a dam of the appropriate genotype; for example:

$$\Pr(AA \text{ x } AA) = \Pr_s(AA)\Pr_d(AA) = 0.25 \text{ x } 0.25 = 0.0625.$$

These probabilities are summarized in Table R2.1.

If a Aa sire is chosen, $\Pr(A_s) = \Pr(a_s) = 0.5$. But if the sire chosen is AA, he produces only A gametes;

$$\Pr(A_s) = 1, \Pr(a_s) = 0.$$

Table R2.1. Random mating among offspring of $Aa \times Aa$

Sire choice	Dam choice	Probability of mating
AA	AA	$0.25 \times 0.25 = 0.0625$
AA	Aa	$0.25 \times 0.5 = 0.125$
AA	aa	$0.25 \times 0.25 = 0.0625$
Aa	AA	$0.5 \times 0.25 = 0.125$
Aa	Aa	$0.5 \times 0.5 = 0.25$
Aa	aa	$0.5 \times 0.25 = 0.125$
aa	AA	$0.25 \times 0.25 = 0.0625$
aa	Aa	$0.25 \times 0.5 = 0.125$
aa	aa	$0.25 \times 0.25 = 0.0625$

Table R2.2 Conditional probabilities for gamete choices

Gamete genotype	Condition (sire genotype)	Conditional probability
A_s	AA_s	$\Pr(A_s \mid AA_s) = 1$
a_s	AA_s	$\Pr(a_s \mid AA_s) = 0$
A_s	Aa_s	$\Pr(A_s \mid Aa_s) = 1/2$
a_s	Aa_s	$\Pr(a_s \mid Aa_s) = 1/2$
A_s	aa_s	$\Pr(A_s \mid aa_s) = 0$
a_s	aa_s	$\Pr(a_s \mid aa_s) = 1$

Sire gamete probabilities are not independent of the choice of the genotype of the sire, but conditional upon that choice. We express this sort of relationship in terms of conditional probabilities:

$$\Pr(A_s : AA_s) = 1,$$

where Pr(A|B) is read as "the probability of A given B".

The complete set of conditional probabilities of sire gamete choices is shown in Table R2.2. Exactly the same conditional probabilities apply, of course, to the dam gamete choices.

Because the choice of gametes controls the probabilities of offspring genotypes, the probability that a random mating will produce offspring of a given genotype is conditional upon the probabilities of the choice of mates. For example,

$$\Pr(Aa_o \mid AA_s \text{ x } Aa_d) = \Pr(A_s \mid AA_s)\Pr(a_d \mid Aa_d) + \Pr(a_s \mid AA_s)\Pr(A_d \mid Aa_d)$$

$$= (1)(0.5) + (0)(0.5) = 0.5.$$

Half the offspring from a AA x Aa mating will be heterozygotes. The complete set of conditional probabilities is given in Table R2.3.

The overall probability of a given offspring genotype is a weighted sum of conditional probabilities (Table R2.3), each being weighted by the probability of the

Table R2.3. Conditional offspring probabilities

Parental Genotypes	*AA*	*Aa*	*aa*
AA × *AA*	1	0	0
AA × *Aa*	0.5	0.5	0
AA × *aa*	0	1	0
Aa × *AA*	0.5	0.5	0
Aa × *Aa*	0.25	0.5	0.25
Aa × *aa*	0	0.5	0.5
aa × *AA*	0	1	0
aa × *Aa*	0	0.5	0.5
aa × *aa*	0	0	1

condition (Table R2.l). For example, the probability of a *AA* offspring is

$$\begin{aligned}\Pr(AA_o) = {} & \Pr(AA_s \times AA_d)\Pr(AA_o|AA_s \mathrm{x} AA_d)\\ & +\Pr(AA_s \mathrm{x} Aa_d)\Pr(AA_o|AA_s \mathrm{x} Aa_d)\\ & +\Pr(AA_s \mathrm{x} aa_d)\Pr(AA_o|AA_s \mathrm{x} aa_d)\\ & +\ldots+\Pr(aa_s \mathrm{x} aa_d)\Pr(AA_o|aa_s \mathrm{x} aa_d)\\ = {} & (0.0625)(1) + (0.125)(0.5) + (0.0625)(0) + \ldots + (0.0625)(0)\\ = {} & 0.25.\end{aligned}$$

A complete set of weighted conditional probabilities, with their frequencies, is given in Table 1.1, Lecture 1. In this table, D represents the probability of choosing a *AA* sire or dam; 2H, a *Aa*; and R, a *aa*. Thus, if we let D = R = 0.25, 2H = 0.5, Table 1.1 would represent the results of random mating among the offspring of a *Aa* × *Aa* mating.

It has been stated that a posteriori estimates of probabilities can be obtained from observed frequencies, and that, if based upon a reasonably large number of trials, these are likely to be close to the true probabilities. In genetics, however, we have an extensive body of knowledge which permits us to predict the probabilities and frequencies a priori. What is more likely to be in question is the validity of the genetic model we have hypothesized to explain the observed phenotypes. We therefore estimate the probabilities with which various phenotypes occur, from the observed frequencies of these phenotypes, then determine whether these estimated probabilities are consistent with the hypothesized genetic mechanism.

Suppose, for example, that we have observed an abnormal phenotype in a large colony of mice, and have noted that it tends to appear in related individuals, so that we are encouraged to believe that it may be of genetic origin. We advance the hypothesis that it is due to a recessive gene, *m*, expressed only in *mm* homozygotes. This is the usual "first guess" hypothesis of the geneticist. Ockam's razor enjoins "Do not multiply entities without necessity", and inheritance of the abnormality by a recessive gene at a single locus is the simplest explanation, invoking the fewest entities, that the geneticist can devise. Only if this hypothesis is palpably not tenable will the geneticist consider more complex models of inheritance.

We will further suppose that abnormal mice are rare, and that those that do appear die before reaching sexual maturity, so that we cannot test our hypothesis by breeding from abnormals. However, under our hypothesis, any mating that yields even a single *mm* offspring must be *Mm* × *Mm*. We expect 25% abnormal offspring from any such mating. We therefore produce as many offspring as we can from each mating that yields abnormal offspring. The greater the number of offspring that we can include in our observed frequency, the better will be our estimate of the underlying probability.

We also seek to increase the number of *Mm* × *Mm* matings available by mating together sibs of the abnormal offspring, again to improve our probability estimate. We know that 2/3 of the normal sibs of a *mm* mouse should be *Mm*, so that nearly half of random matings among such sibs should be heterozygote × heterozygote. We observe the offspring of these matings to choose those that produce abnormal offspring.

Over some period of time, then, we produce a reasonable number of offspring, say N, from matings that must be *Mm* × *Mm* if our hypothesis is correct. Among these offspring, n_1 are normal and n_2 are abnormal, $n_1 + n_2 = N$. The frequency of abnormal offspring observed, n_2/N, is our estimate of the probability of an abnormal phenotype. From the a priori probabilities under our hypothesis we predict this probability to be 0.25.

We will reject the hypothesis if n_2/N is significantly different from 0.25. But because our sample of offspring is finite, n_2 is subject to sampling error, and any observed difference between n_2/N and 0.25 may be due to that cause. We use the χ^2 ratio test to determine whether the deviation of n_2/N from 0.25 is attributable to sampling.

We calculate a χ^2 value according to the formula

$$\chi^2 = (n_1 - e_1)^2/e_1 + (n_2 - e_2)^2/e_2,$$

where e_1 and e_2 are the expected numbers of normal and abnormal offspring if our hypothesis is correct. (In this case, $e_1 = 3N/4$ and $e_2 = N/4$.) We then compare the calculated χ^2 value to a table of the χ^2 statistic. The table shows the probability that a χ^2 value equal to or greater than that calculated from the sample can arise by chance when the hypothesized mode of inheritance is true. If this probability is high, we can conclude that the difference between n_i and e_i is due to sampling error, and that the hypothesized mode of inheritance is a satisfactory explanation of the origin of the abnormality. But if the probability that the difference arose by chance is low, we must conclude that the hypothesized mode of inheritance is not a satisfactory explanation, and proceed to elaborate a more complex model to explain the condition.

As in all cases of hypothesis testing, there is some probability that we will make the wrong decision. The inheritance model may actually be correct, but we may reject it because we happen to have drawn a sample in which there are too many or too few abnormal offspring. The χ^2 we have calculated has a low probability of occurring; but it still can occur by chance. This type of wrong decision is called a Type I error.

On the other hand, the hypothesis may be wrong. If the observed number of abnormal offspring is not very different from the number expected under the incorrect hypothesis, the calculated χ^2 value has a high probability of occurring under the incorrect hypothesis. Therefore, we accept that hypothesis. This is called a Type II error. It is particularly likely to occur if the true model of inheritance gives a priori probabilities which are not very different from those given by our simple recessive hypothesis.

We can control the amount of Type I error by choosing the probability level at which we will reject the hypothesis. If we decide to reject the hypothesis whenever the calculated χ^2 value has a probability of, say, 5% or less, the risk of Type I error is 5%. By extending the rejection zone to 1% or less, we reduce the risk of Type I error to only 1%.

Unfortunately, when we reduce the probability of a Type I error, we increase that of a Type II error, and vice versa. We cannot predetermine the risk of Type II error we incur when we choose a given level of Type I error risk, as that depends on the a priori probabilities inherent in the true model of inheritance.

The compromise that we make between risks of Type I and Type II error depends principally on the penalties involved for making each type of wrong decision. Will our reputations as geneticists suffer more if we attribute abnormalities to simple inheritance models when they are actually under complex control? Or if we claim some unusual and complicated model of inheritance for a simple recessive trait? It can be depended upon that, sooner or later, others will reveal our errors by deducing the true nature of the inheritance of the abnormality.

Review Lecture 3 Statistical Methods

When we measure the amount of some characteristic possessed or produced by each member of a group of organisms, we may be interested only in that particular set of measurements. If so, once we have a measurement, an X value, for each member of the group, we can rest content. We can calculate an average value, measure the range of values, graph the distribution, etc. The results are not estimates of the average, range, or distribution; they are the actual values for that set of measurements.

Almost always, however, we will not be interested in the set of X's we have measured as such. We will want to use them as representative of some larger population of X's. We want to know something about the distribution, not of the sample of X values that we have measured, but of the underlying population from which that sample was drawn. We must, therefore, make statistical inferences about the population on the basis of the sample.

First of all, to make valid inferences, we must have a random sample from the population. That is, our sample must be chosen in such a way that every member of the population has equal probability of appearing in the sample. In the ensuing discussion, we will assume that we have a random sample of n X values from a population of N.

X is a variable, and both the X's in our sample and those in the population as a whole are distributed along some scale of values according to some unknown law. If we wish to characterize the population distribution fully, we must have a very large sample, n nearly as large as N; and this is seldom possible. Fortunately, for most purposes we do not need to characterize the population fully, but merely to have reasonable estimates of (1) the location of the center of the population on the value scale, and (2) the extent to which the population spreads along the scale.

The most commonly used measure of the center of the distribution is the arithmetic mean. The arithmetic mean, μ, of the population is defined as the center of gravity of the distribution of X's. Some X values will be greater than μ, some less; if we were to take the deviation of each X from μ, some would be positive, some negative. Over the whole population, the total of the positive deviations will exactly balance the total of the negative deviations, so that

$$\Sigma(X_i - \mu) = 0. \qquad \text{(R3.1)}$$

In a given population, there might be more positive than negative deviations; if so, the negative deviations would tend to be slightly larger so that the two sums are equal.

Of course, we do not know what μ is for the population that we have sampled, but we can estimate it. The best estimate of μ from the sample is $\bar{x}$, which is the sum of the sample X values, divided by the number of values in the sample:

$$\bar{x} = \Sigma X_i / n. \tag{R3.2}$$

As μ is the center of gravity of the population, $\bar{x}$ is the center of gravity of the sample;

$$\Sigma(X_i - \bar{x}) = 0$$

in the sample.

As an estimate of μ, $\bar{x}$ has a number of useful qualities, of which we will only mention that it is unbiased. If we were to take an infinite number of samples of size n from the population, and calculate $\bar{x}$ from each, we would create a population of $\bar{x}$'s. The mean of that population of sample means would be μ.

Under special circumstances, we may be interested in two other mean values. The use of $\bar{x}$ implies a linear model of X, with the value of any particular X being the sum of the effects of a number of factors. If X is more correctly regarded as the product of the effects of the factors, the geometric mean, G*, may be more appropriate:

$$G^* = (\Pi X_i)^{1/n}.$$

This sounds formidable, but the geometric mean is merely the antilogarithm of the arithmetic mean of the logarithms of the X values:

$$\log G^* = \Sigma(\log X_i)/n.$$

The harmonic mean, H*, may be appropriate as a measure if X is a function of the reciprocals of the factor effects. H* is the reciprocal of the arithmetic mean of the reciprocals of the X values:

$$H^* = [\Sigma(1/X_i)/n]^{-1}.$$

As with the arithmetic mean, the geometric and harmonic means are estimated by taking the corresponding function of the sample values. G* and H* are also means, but the word "mean" by itself can usually be taken to refer to the arithmetic mean, and will be so interpreted in the present text.

Two other measures of location deserve mention: the median and the mode. The median is the central value of a sample. If we were to rank all the X's in the sample from lowest to highest, the median would be the value in the middle of the ranking, n/2 values from either extreme. The mode is the most common value in the distribution. Over all the X's, we expect repetitions of the same value, especially those values toward the middle of the distribution. The mode is the value which is repeated most often.

If we have a symmetrical distribution with a single central peak, mean, median and mode will all be the same value. But if the distribution is not symmetrical, if, say, more X values lie below than above the mean, the mode will be the least of the three values, the median next, and the mean the largest. If the distribution has more than one peak, there may be several modes; if it has no peak, as in the case of a rectangular distribution, for example, there will be no mode at all; or perhaps it would be more correct to say, every value will be modal.

Median and mode have been used traditionally for certain distributions, especially in the social sciences, and have proven useful. Both suffer from one serious fault, however; the sample mode and median do not in general provide unbiased estimators of the mode and median of the population. In fact, if we do not know the nature of the distribution, the relationships of these sample statistics to the corresponding population parameters are undefined. The sample mean, on the other hand, is an unbiased estimator of the population mean, whatever the underlying distribution.

A measure of the spread or dispersion of the population of X values is necessary in two situations. If we wish to estimate the population mean, our best guess is $\bar{x}$; but we know that unless $n=N$, $\bar{x}$ is not likely to be exactly equal to μ. With a measure of the spread of the distribution, we can estimate a range of values around $\bar{x}$ within which we can expect the population mean to lie.

Alternatively, we may wish to determine whether the means of two populations are the same. We can compare the means of samples from the two populations; but if they differ, they may do so because the population means differ, or merely because neither sample mean is a perfect estimator of the mean of its population. A measure of spread in the population can assist us in deciding whether the sample means reflect actual differences in the means of the populations, or merely differ by chance.

There are numerous measures of spread or dispersion that can be employed, but the one most commonly used is the variance, σ^2. This is defined as the average squared deviation of the X values from their mean,

$$\sigma^2=\Sigma(X_i-)^2/N. \tag{R3.3}$$

We have already seen that the sum of the deviations from the mean across the whole population is 0, plus and minus deviations balancing one another. If we square each deviation, however, the square is always positive, whether the deviation itself is positive or negative. Therefore, $\Sigma(X_i-\mu)^2$ is always a positive quantity.

The corresponding function of the sample X values,

$$\Sigma(X_i-\bar{x})^2/n,$$

provides a biased estimate of σ^2. To obtain an unbiased estimate, we must divide the sum of the squared deviations in the sample by $n-1$:

$$s^2=\Sigma(X_i-\bar{x})^2/(n-1). \tag{R3.4}$$

The appropriate divisor for the sample variance is $n-1$, rather than n, because we have used one degree of freedom from the sample to estimate $\bar{x}$.

The appropriate function of the estimated population variance, subtracted from $\bar{x}$, gives the lower limit of a confidence interval within which the population mean can be expected to lie; the same function, added to $\bar{x}$, gives the upper limit. The function used depends on the desired level of confidence in the result. If, for example, we desire to be 95% sure that μ lies within the interval, we can use $2[\mathrm{Var}(X)/n]^{1/2}$ as the function of the sample variance. The higher the desired level of confidence, of course, the wider the interval must be made. The above function of the variance estimate is only valid if the population is a Gaussian "normal" distribution.

Comparing means of populations is referred to as hypothesis testing, because the comparison is made by hypothesizing that the means actually do not differ (even if the experimenter believes that they do). This null hypothesis is then tested by the appropriate procedure. A simple comparison of two means can be made using Student's t test. The t statistic, multiplied by a function of the estimated variance, measures the largest difference between two means that could be expected to occur by chance, if the populations being compared actually have the same mean. If the difference between the two measured $\bar{x}$ values is greater than this, we can reject the null hypothesis, concluding that the population means actually differ. If the difference between sample means is less than the t test value, we conclude that the difference is due merely to sampling error; that the two populations have the same mean or that the two samples actually came from the same population. Again, the alternate risks of Type I and Type II error are present (see Review Lecture 2).

A more powerful and more general means of testing hypotheses lies in the analysis of variance procedure. Suppose, for example, that we want to know whether different genotypes at the brown pigment locus in mice cause differences in body weight. This amounts to asking whether the mean weights of populations of *BB*, *Bb*, and *bb* mice differ.

We sample at random n individuals from each of the three genotypes. For simplicity, we will assume that the samples are uniform, differing only in B-locus genotypes. We weigh each individual, and calculate the sample mean for each genotype. Almost certainly, the three sample means will differ to some extent, and we want to know whether the differences reflect real differences among the three genotypic populations, or whether they have arisen due to sampling error.

Our null hypothesis is that the population means do not differ. We then calculate from the data the total variance among all observations, the variance among the sample mean values for the three genotypes, and the average variance within the samples from each genotype. Formulae for calculating the sum of squares associated with each of these sources of variation are given in Table R3.1. Dividing each sum of squares by the associated degrees of freedom (df) gives a mean square, which is the estimated variance, for each source of variation.

In Table R3.1, we have symbolized the number of groups being compared by t, with n observations per group. In the case of the B-locus genotypes, t=3, of course. Also in the table, we have used the dot notation for sums taken prior to squaring; if, for example, we sum the observations within the i-th group, square

Table R3.1. The analysis of variance

Source of variation	df	SSq	Ex(MSq)
Total	$tn-1$	$\Sigma(X_{ij}^2)-X_{..}^2/tn$	$\sigma_W^2+[n(t-1)/(tn-1)]\sigma_B^2$
Between groups	$t-1$	$\Sigma(X_{i.}^2)/n-X_{..}^2/tn$	$\sigma_W^2+n\sigma_B^2$
Within groups	$t(n-1)$	$\Sigma[\Sigma(X_{ij}^2)-\Sigma(X_{i.}^2)/n]$	σ_W^2

that sum, and total the squares over all groups, this would be represented as $\Sigma(X_{i.})^2$.

The mean squares in the analysis represent various functions of the variance within groups, $\sigma_W{}^2$, and that between groups, $\sigma_B{}^2$. These functions are shown in Table R3.1.

Under the null hypothesis that the population means are not different, $\sigma_B{}^2=0$, and therefore MSq(Between) and MSq(Within) are independent estimates of $\sigma_W{}^2$. Then the ratio MSq(Between)/MSq(Within) should be distributed as the F statistic. If the null hypothesis is wrong, the ratio of the mean squares should be greater than F, since $\sigma_B{}^2$ must be positive if it exists at all. The calculated ratio s compared to tabled values of F. The probability that an F value as large as or larger than the calculated ratio can occur by chance alone can then be read from the F table. This probability sets the risk of Type I error. Suppose that we decide to accept a 5% risk of Type I error; then, if the calculated F does not exceed the tabled F for that risk, we conclude that the null hypothesis is correct, and that the B-locus genotypes indeed do not differ in body weight. But if the calculated F exceeds the tabled F, we reject the null hypothesis, concluding that the population means for the three genotypes are different.

Adaptations of this basic analysis can be used to test hypotheses with regard to more complex relationships. For example, if half of the n mice sampled from each genotype were male and half female, we could also test for sex differences in body weight, and for interactions between genotype and sex. That is, we could test whether males and females differed, and whether males of different genotypes differed while females did not. The details of this and other more complex analyses must be left to texts on experimental design and analysis (e.g., Federer 1955).

The analysis of variance procedure can also be used to estimate components of variance, such as $\sigma_B{}^2$ and $\sigma_W{}^2$ in the simple analysis outlined in Table R3.1. The assumptions about the nature of the populations sampled are somewhat different for this procedure to be valid; again, the details will be left to texts in the appropriate area. In the analysis of the inheritance of quantitative traits, we will employ this aspect of the analysis of variance extensively.

If we measure two variables, say X and Y, on the same set of individuals, the fact that X and Y are measured on the same individuals does not mean that X and Y are necessarily related. We can plot the sample values of X and Y on a set of bivariate coordinates, as in Fig. R3.1. In Fig. R3.1a, there is no particular association between the X and Y variables. High X values are associated with high Y values about as often as with low, and the same is true of low X values. In Fig. R3.1b, on the other hand, high X's occur with high Y's more frequently than with low, and there is a similar association of low X's with low Y's. In Fig. R3.1c, this relationship is reversed; high X's are associated with low Y's, and low X's with high Y's, more frequently than not. Fig. R3.1a represents independence of X and Y; b and c, respectively, a positive and a negative relationship between the two variables.

We measure the relationship between X and Y in terms of their covariance, σ_{XY}. Our estimate of this parameter is the sample covariance,

$$\mathrm{Covar}(XY)=\mathrm{SCP}(XY)/(n-1)=[\Sigma X_iY_i-(\Sigma X_i)(\Sigma Y_i)/n]/(n-1), \qquad (R3.5)$$

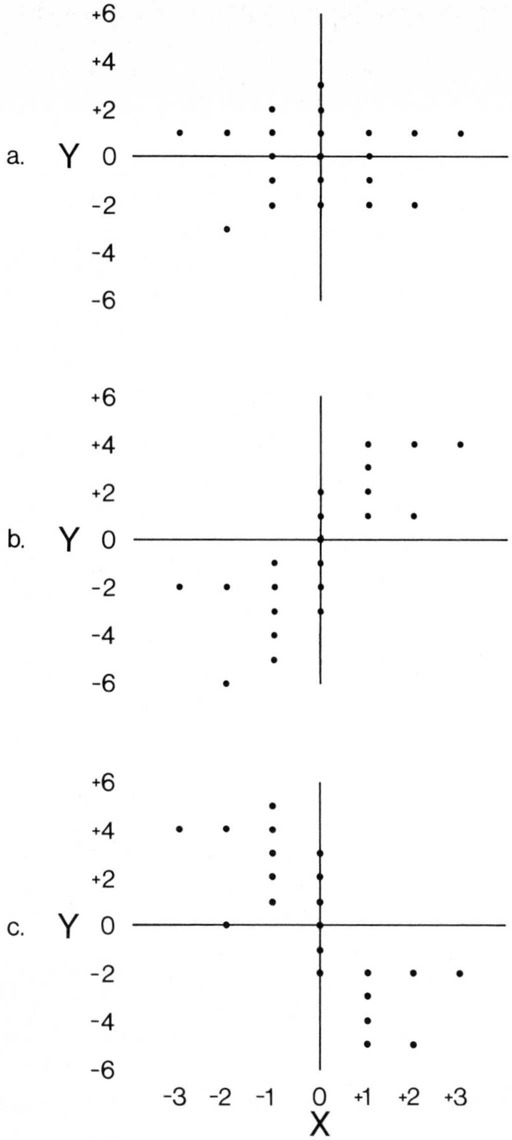

Fig. R3.1. Covariance. (a) If variables X and Y vary independently, +X values occur as often with +Y as with −Y, and vice versa. but if they vary together, either (b) +X with +Y and −X with −Y combinations outnumber the +X with −Y and −X with +Y combinations, in which case there is positive covariance; or (c) +X, −Y and −X, +Y combinations are more common (negative covariance)

where X_i and Y_i are the values of the two variables measured on individual i. Note that the formula is very similar to Eq. (R3.4), the formula for Var(X), except that the cross products of X and Y are substituted for the squares of the X's.

Covar(XY) will be near zero if X and Y are distributed independently, as in Fig. R3.1a. Positive values of X are associated as often with negative as with positive values of Y, and the same is true for negative X values. The crossproducts of (+X)(+Y) or of (−X)(−Y) will be positive; of (+X)(−Y) or of (−X)(+Y), negative. Thus, positive and negative crossproducts balance each other, and SCP(XY) will be zero.

But if +X, +Y and −X, −Y combinations exceed those where the two variables have opposite signs, as in Fig. R3.1b, there will be more positive than negative crossproducts, and the sum will be positive. Similarly, if combinations with opposite signs are in excess (Fig. R3.1c), the sum of crossproducts will be negative.

We measure the strength and direction of the association between X and Y by the correlation coefficient, ϱ. The sample estimate of this parameter is

$$r = [\mathrm{Covar}(XY)^2/\mathrm{Var}(X)\mathrm{Var}(Y)]^{1/2}. \quad \text{(R3.6)}$$

This parameter measures the association of variation in one variable with variation in the other. In fact, r^2 can be used to estimate the proportion of the variance in X that is due to variation in Y. Variation in one variable cannot account for more than 100% of the variation in the other, so ϱ^2 cannot be greater than 1. Thus, ϱ cannot be greater than 1 or less than −1. Estimates of r greater than 1 or less than −1 are occasionally obtained, when estimated indirectly, as from a variance component analysis. They can be interpreted as being due to poor estimates of the variances and covariances involved.

Care must be taken to avoid interpreting correlation as tantamount to causation. Correlation may indicate causation; that is, changes in one variable may cause changes in the other. But correlation may also occur because both variables are controlled by a third factor. Health professionals often cite the positive correlation between number of visits to doctors and mortality. If correlation always implied causation, the conclusion from this statistic might be that "visiting doctors is dangerous to your health." In fact, of course, some illness or injury is the third factor causing both doctor visits and death.

Correlations can also arise purely by chance. The classic case is the correlation between monthly precipitation in the city of New York and the number of r's in the name of the month. It is sheer happenstance that the rainy, snowy winter months happen to have r's in their names, while the drier summer months do not.

Where there is distinct causation of variation in Y due to variation in X, we may be interested in calculating the regression of Y on X. This statistic again involves the covariance, the regression coefficient, β_{YX}, being estimated as

$$b_{YX} = \mathrm{Covar}(XY)/\mathrm{Var}(X). \quad \text{(R3.7)}$$

A nonzero correlation coefficient tells us that X and Y tend to move together, and in which direction. The regression coefficient tells us how much Y changes per unit change in X. If we have estimates of $\bar{x}$, $\bar{y}$, and b_{YX}, we can construct a regression line, which passes through the point $\bar{x}$, $\bar{y}$ and has slope b_{YX}. I.e., the Y value of a point on the regression line with X value $\bar{x}+z$ is $b_{YX}(\bar{y}+z)$. The regression coefficient will have the same sign as the correlation coefficient, and as Covar(XY).

For the regression model to be strictly valid, variation in Y must be caused by variation in X, and X values must be measured without error. For each X value, there is a distribution of Y's, the mean of which is $\hat{Y}$, the Y value on the true regression line. We can estimate these means by the formula

$$\hat{Y}_i - \bar{y} = b_{YX}(X_i - \bar{x}). \quad \text{(R3.8)}$$

$\hat{Y}_i$ in Eq. (R3.8) is our best prediction of the Y associated with X_i. To illustrate the difference between a regression and a correlation situation, suppose that we have body weights on a group of mice weighed at various ages between, say, 21 and 42 days of age. The mice are growing during this period, so that increases in age (X) will cause increases in body weight (Y). We can measure age virtually without error; weight measurements are subject to error. We can therefore validly estimate the regression of weight on age, and can predict the weight of a mouse at any given age (within the age limits of the study) from the regression equation.

Suppose, however, that we also measure tail length on these mice. Tail length will also tend to increase with increasing age, and will therefore be correlated with body weight. But tail length increases do not cause increases in body weight, nor vice versa. Both are caused by increases in age. Neither tail length (Z) nor body weight (Y) can be measured without error, and there is no true independent-dependent variable relationship between them. Therefore, we can estimate a correlation coefficient, but cannot validly estimate a regression coefficient, between Y and Z.

We do not need to have different variables measured on the same individual to produce a covariance. We can also discuss the covariance of the same variable measured on different individuals. We will make considerable use of the covariance between measurements on the same trait in related individuals in our discussion of the inheritance of quantitative traits. In fact, the analysis of variance is a covariance procedure. Individuals are classified into groups. The variance among group means is the within group covariance.

References

Allison AC (1954) Notes on sickle cell polymorphism. Ann Hum Genet 19:39–51

Bateson W, Punnett RC (1906) Gametic coupling. Rep Evol Comm R Soc London 3:8–111

Bell AE (1977) Heritability in retrospect. J Hered 68:297–300

Bell AE (1982) The *Tribolium* model in animal breeding research. Proc 2nd World Congr Genet Appl Livestock Production. Garsi, Madrid, pp 26–42

Bennett JH (1963) Random mating and sex linkage. J Theor Biol 4:28–36

Bernstein F (1924) Ergebnisse einer biostatistischen zusammenfassenden Betrachtung über die erblichen Blutstrukturen des Menschen. Klin Wochenschr 3, II:1495–1497

Bernstein F (1925) Zusammenfassende Betrachtungen über die erblichen Blutstrukturen des Menschen. Z Indukt Abstamm Vererbungsl 37:237–270

Blakeslee AF, Salmon MR (1931) Odor and taste blindness. Eugen News 16:105–109

Bridges CB (1913) Non-disjunction of the sex chromosomes of Drosophila. J Exp Zool 15:587–606

Bridges CB (1914) Direct proof through non-disjunction that the sex-linked genes of *Drosophila* are borne by the X-chromosome. Science 40:107–109

Bridges CB (1916) Non-disjunction as proof of the chromosome theory of heredity. Genetics 1:1–52, 107–163

Bruce AB (1910) The Mendelian theory of heredity and the augmentation of vigor. Science 32:627–628

Castle WE (1903) The laws of heredity of Galton and Mendel and some laws governing race improvement by selection. Proc Am Acad Arts Sci 39:223–242

Castle WE, Little CC (1910) On a modified Mendelian ratio among yellow mice. Science 32:868–870

Cepellini R, Siniscalco M, Smith CAB (1955) The estimation of gene frequencies in a random-mating population. Ann Hum Genet 20:97–115

Cloudman AM, Bunker LE (1945) The varitint-waddler mouse. A dominant mutation in *Mus musculus*. J Hered 36:259–263

Collins GN (1921) Dominance and the vigor of first generation hybrids. Am Nat 55:116–133

Correns C (1900) G. Mendel's Regel über das Verhalten der Nachkommenschaft der Rassenbastarde. Ber Dtsch Bot Ges 18:158–168 [Transl. Genetics (Suppl) 35:33–41, 1950]

Correns C (1902) Über den Modus und den Zeitpunkt der Spaltung der Anlagen bei den Bastarden vom Ebsen-Typus. Botan Z 60, II, 5/6:66–82

Creighton HB, McClintock B (1931) A correlation of cytological and genetical crossing over in *Zea mays*. Proc Natl Acad Sci USA 17:492–497

Crow JF (1948) Alternative hypotheses of hybrid vigor. Genetics 33:477–487

Crow JF (1954) Breeding structure of populations. III. Effective population number. In: Kempthorne O, Bancroft TA, Gowen JW, Lush JL (eds) Statistics and mathematics in biology. Iowa State College Press, Ames

Crow JF (1983) Genetics notes: an introduction to genetics, 8th edn. Burgess, Minneapolis

Crow JF (1986) Basic concepts in population, quantitative, and evolutionary genetics. Freeman, New York

Crow JF, Kimura M (1970) An introduction to population genetics theory. Harper & Row, New York

Cruden D (1949) The computation of inbreeding coefficients for closed populations. J Hered 40:248–251

Cuenot L (1905) Les races pures et leurs combinaisons chez les souris (4th note). Arch Zool Exp Gen (4th ser) 3:cxxiii-cxxxii

Darwin C (1859) The origin of species. Modern Library edition, Random House, New York

Dawson PS, Riddle RA (1975) A genetic maternal effect on egg surface in *Tribolium castaneum*. J Hered 66:31–32

Dobzhansky Th (1941) Genetics and the origin of species. Columbia Univ Press, New York

Doncaster L, Raynor GH (1906) On breeding experiments with Lepidoptera. Proc Zool Soc London 1:125–133

Doolittle DP (1961) The effect of single gene substitutions on resistance to radiation in mice. Genetics 46:1501–1509

Doolittle DP (1968) An extension of Snyder's ratios to the case of epistasis. J Hered 59:117–119

Doolittle DP (1979) Selection for spot size in mice. J Hered 70:390–394

Doolittle DP (1981) Effects of selection for spot size on reproduction and body weight in mice. Theor Appl Genet 60:329–331

Doolittle DP, Wilson SP (1979) A comparison between terminal and rotational three-way crosses in mice. J Anim Sci 48:280–285

Dronamraju KR (1965) The function of the Y-chromosome in man, animals and plants. Adv Genet 13:227–310

Durham FM, Marryat DCE (1908) Note on the inheritance of sex in canaries. Rep Evol Comm R Soc London 4:57–60

East EM (1908) Practical use of mendelism in corn breeding. Inbreeding in corn. Conn Agric Exp Stn New Haven Rep 1907, pp 406–418

East EM (1910) A Mendelian interpretation of variation that is apparently continuous. Am Nat 44:65–82

East EM (1935) Genetic reactions in *Nicotiana*. III. Dominance. Genetics 20:443–451

Emerson RA, East EM (1913) The inheritance of quantitative characters in maize. Nebr Agric Exp Stn Res Bull 2

Emik LO, Terrill CE (1949) Systematic procedures for calculating inbreeding coefficients. J Hered 40:51–55

Fabricius-Hansen V (1940) Blood groups and MN types of Eskimos. II. (District of Julianehaab, West Greenland.) J Immunol 38:405–411

Falconer DS (1960) Introduction to quantitative genetics. Oliver & Boyd, Edinburgh

Falconer DS (1981) Introduction to quantitative genetics, 2nd edn. Longman, London

Falk CT, Li CC (1969) Negative assortative mating: exact solution to a simple model. Genetics 62:215–223

Federer WT (1955) Experimental design, theory and applications. Macmillan, New York

Finney DJ (1952) The equilibrium of a self-incompatible polymorphic species. Genetics 26:33–64

Fisher RA (1918) The correlation between relatives on the supposition of Mendelian inheritance. Trans R Soc Edinburgh 52:399–433

Fisher RA (1930) The genetical theory of natural selection. Clarendon Press, Oxford

Fisher RA (1934) The effect of method of ascertainment upon the estimation of frequencies. Ann Eugen 6:13–25

Fitch WW, Atchley WR (1985a) Evolution in inbred strains of mice appears rapid. Science 228: 1169–1175

Fitch WW, Atchley WR (1985b) Rapid mutation in mice? Science 230:1408–1409

Ford CE, Jones K, Polani P, Almeida de J, Briggs J (1959) A sex chromosome anomaly in a case of gonadal dysgenesis (Turner's syndrome). Lancet 1:711–713

Fox AL (1931) Science news items–tasteblindness. Science 73, no 1894 (Suppl.):14

Garrod AE (1909) Inborn errors of metabolism. Frowde, London

Goodale HD (1918) Internal factors influencing egg production in the Rhode Island Red breed of domestic fowl. Am Nat 52:65–94, 209–232, 301–321

Goodale HD, McMullen G (1919) The bearing of ratios on theories of the inheritance of winter egg production. J Exp Zool 28:83–124

Goodale HD, Sanborn R (1922) Changes in egg production in the station flock. Mass Agric Exp Stn Bull 211

Green EL (1981) Genetics and probability in animal breeding experiments. Oxford Univ Press, New York

Green EL, Doolittle DP (1963) Systems of mating used in mammalian genetics. In: Burdette WJ (ed) Methodology in mammalian genetics. Holden-Day, San Francisco

Green MC (1966) Mutant genes and linkages. In: Green EL (ed) Biology of the laboratory mouse, 2nd edn., Chapter 8. McGraw-Hill, New York

Green MC, Bailey DW, Green EL, Roderick TH, Russell ES (1985) Rapid mutation in mice? Science 230: 1407–1408

Haldane JBS (1930) Theoretical genetics of polyploids. J Genet 22:359–372

Hardy GH (1908) Mendelian proportions in a mixed population. Science 28:49–50

Hays FA (1924) The application of genetic principles in breeding poultry for egg production. Poult Sci 4:43–50

Hazel LN, Lush JL (1950) Computing inbreeding and relationship coefficients from punched cards. J Hered 41:301–306

Hill WG (1972a) Estimation of realized heritabilities from selection experiments. I. Divergent selection. Biometrics 28:747–766

Hill WG (1972b) Estimation of realized heritabilities from selection experiments. II. Selection in one direction. Biometrics 28:767–780

Hill WG (1972c) Effective size of populations with overlapping generations. Theor Popul Biol 3:278–289

Hill WG (1979) A note on effective population size with overlapping generations. Genetics 92:317–322

Hill WG (1985) Quantitative Genetics. Part I. Explanation and analysis of continuous variation. Part II. Selection. Benchmark Pap in Genet Ser. Van Nostrand, New York

Hubby JL, Lewontin RC (1966) A molecular approach to the study of genic heterozygosity in natural populations. I. The number of alleles at different loci in *Drosophila psuedoobscura.* Genetics 54:577–594

Hulbert LL, Doolittle DP (1971) Abnormal skin and hair: a sex-linked mutation in the house mouse. Genetics 68:s29

Hulbert LL, Doolittle DP (1973) Inheritance of white spotting in mice. J Hered 64:260–264

Huskins CL (1931) The origin of *Spartina townsendii.* Genetica (The Hague) 12:531–538

Hutt FB (1949) Genetics of the fowl. McGraw-Hill, New York

Ibsen HL, Steigleder E (1917) Evidence for the death in utero of the homozygous yellow mouse. Am Nat 51:740–752

Jacobs PA, Strong JA (1959) A case of human intersexuality having a possible XXY sex-determining mechanism. Nature (London) 183:302–303

Janssens FA (1909) La théorie de la chiasmatypie. Cellule 25:389–411

Jennings HS (1916) The numerical results of diverse systems of breeding. Genetics 1:53–89

Jennings HS (1917) The numerical results of diverse systems of breeding, with respect to two pairs of characters, linked or independent, with special relation to linkage. Genetics 2:97–154

Johanssen W (1903) Über Erblichkeit in Populationen und in seinen Linien. Fischer, Jena

Johnson FM, Aquadro CF, Skow LC, Langley CH, Lewis SE (1985) Rapid mutation in mice? Science 230:1406–1407

Jones DF (1918) The effects of inbreeding and crossbreeding upon development. Conn Agric Exp Stn Bull 207

Karpechenko GD (1927) The production of polyploid gametes in hybrids. Hereditas 9:349–368

Karpechenko GD (1928) Polyploid hybrids of *Raphanus sativa* L. x *Brassica oleracea* L. Z Indukt Abstamm Vererbungsl 48:1–85

Keeble F, Pellew C (1910) The mode of inheritance of stature and of time of flowering in peas (*Pisum sativum*). J Genet 1:47–56

Kimura M (1955) Solution of a process of random genetic drift with a continuous model. Proc Natl Acad Sci USA 41:144–150

Kimura M (1983) The neutral theory of molecular evolution. Camb Univ Press

King RC, Stansfield WD (1985) A dictionary of genetics, 3rd edn. Oxford Univ Press, New York

King SC, Henderson CR (1954) Variance components analysis in heritability studies. Poult Sci 33:147–154

King SC, Carson JR, Doolittle DP (1959) The Connecticut and Cornell randombred populations of chickens. World's Poult Sci J 15:139–159

Klinefelter HF, Reifenstein EC, Albright F (1942) Gynecomastia, aspermatogenesis without aleydigism and increased secretion of follicle-stimulating hormone. J Clin Endocrinol 2:615–627

Kusakabe S, Mukai T (1984) The genetic structure of natural populations of *Drosophila melanogaster*. XVII. A population carrying genetic variability explicable by the classical hypothesis. Genetics 108:393–408

Lenz W (1961) Medical genetics. Chicago Univ Press
Lerner IM (1950) Population genetics and animal improvement. Camb Univ Press
Lerner IM (1954) Genetic homeostasis. Oliver & Boyd, London
Li CC (1955) Population genetics. Chicago Univ Press
Li CC (1967a) Castle's early work on selection and equilibrium. Am J Hum Genet 19:70–74
Li CC (1967b) Genetic equilibrium under selection. Biometrics 23:397–484
Li CC (1975) Path analysis – a primer. Boxwood Press, Pacific Grove, Cal
Li CC (1976) First course in population genetics. Boxwood Press, Pacific Grove, Cal
Little CC (1916) The occurrence of three recognised color mutations in mice. Am Nat 50:335–349
Lush JL (1936) Genetic aspects of the Danish system of progeny-testing swine. Iowa Agric Exp Stn Res Bull 204:105–195
Lush JL (1940) Intra-sire correlations or regressions of offspring on dam as a method of estimating heritability of characteristics. Proc Am Soc Anim Prod 33:293–301
Lyon MF (1961) Gene action in the X-chromosome of the mouse (*Mus musculus*, L.). Nature (London) 190:372–373
Lyon MF (1962) Sex chromatin and gene action in the mammalian X-chromosome. Am J Hum Genet 14:135–148
MacArthur JW (1949) Selection for small and large body size in the house mouse. Genetics 34:194–209
Malecot G (1948) Les mathématiques de l'hérédité. Masson, Paris
Mather PL (1979) Factors affecting heritability estimates of pre-school-aged twins' language performance skills. Thes Purdue Univ, Lafayette, Ind
Mendel G (1865) Experiments in plant hybridization. Verh Naturforsch Ver Brünn, Abhandl iv
Mertens TR (1971) Speciation in *Spartina*: a classical example erroneously reported in text books. BioSci 21:420–421
Morgan TH (1910a) Chromosomes and heredity. Am Nat 44:449–496
Morgan TH (1910b) Sex limited inheritance in Drosophila. Science 32:120–122
Morgan TH, Sturtevant AH, Muller HJ, Bridges CB (1915) The mechanism of mendelian heredity. Holt, New York
Morris D (1967) The naked ape. Cope, London
Mukai T (1964) The genetic structure of natural populations of *Drosophila melanogaster*. I. Spontaneous mutation rate of polygenes controlling viability. Genetics 50:1–19
Pearson K (1904) Mathematical contributions to the theory of evolution. XII. On a generalised theory of alternative inheritance, with special reference to Mendel's laws. Phil Trans R Soc London Ser A 203:53–86
Plum M (1954) Computation of inbreeding and relationship coefficients. J Hered 45:92–94
Race RR (1944) An "incomplete" antibody in human serum. Nature (London) 53:771–772
Robbins RB (1918) Some applications of mathematics to breeding problems III. Genetics 3:375–389.
Shull GH (1911) The genotypes of maize. Am Nat 45:234–252
Shull GH (1914) Duplicate genes for capsule form in *Bursa bursa-pastoris*. Z Indukt Abstamm Vererbungsl 12:97–149
Silvers W (1979) The coat colors of mice. Springer, Berlin Heidelberg New York
Sinnott EW, Dunn LC, Dobzhansky Th (1950) Principles of genetics, 4th edn. McGraw-Hill, New York
Sinnott EW, Dunn LC, Dobzhansky Th (1958) Principles of genetics, 5th edn. McGraw-Hill, New York
Smith CAB (1956) A test for segregation ratios in family data. Ann Hum Genet 20:257–265
Snyder LH (1931) Inherited taste deficiency. Science 74:151–152
Snyder LH (1932) Studies in human inheritance. IX. The inheritance of taste deficiency in man. Ohio J Sci 32:436–440
Srb AM, Owen RD, Edgar RS (1965) General genetics, 2nd edn. Freeman, San Francisco
Stebbins GL (1950) Variation and evolution in plants. Columbia Univ Press, New York
Stern C (1926) Vererbung im Y-Chromosom von *Drosophila melanogaster*. Biol Zentralbl 46:344–348
Stern C (1927) Ein genetischer und zytologischer Beweis fur Vererbung im Y-Chromosom von *Drosophila melanogaster*. Z Indukt Abstamm Vererbungsl 44:187–231
Stern C (1943) The Hardy-Weinberg law. Science 97:137–138
Stern C (1957) The problem of complete Y-linkage in man. Am J Hum Genet 9:147–166

Stern C (1962) Wilhelm Weinberg, 1862–1937. Genetics 47:1–5
Sturtevant AH (1965) A history of genetics. Harper & Row, New York
Sutton WS (1902) The chromosomes in heredity. Biol Bull Marine biol Lab 4:231–248
Thompson JS, Thompson MW (1980) Genetics in medicine, 3rd edn. Saunders, Philadelphia
Truesdale-Mahoney N, Doolittle DP, Weiner H (1981) Genetic basis for the polymorphism of rat liver cytosolic aldehyde dehydrogenase. Biochem Genet 19:1275–1282
Tschermak E (1900) Über kunstliche Kreuzung bei *Pisum sativum*. Ber Dtsch Bot Ges 18:232–239 [Transl. Genetics (Suppl) 35:42–47]
Turner HH (1938) A syndrome of infantilism, congenital webbed neck and cubitus valgus. Endocrinology 23:566–574
Tyzzer EE, Little CC (1916) Studies on the inheritance of susceptibility to a transplantable sarcoma (J.W.B.) of the Japanese waltzing mouse. J Cancer Res 1:387–389
Vries de H (1900) Sur la loi de disjonction des hybrides. C R Acad Sci (Paris) 130:845–847 [Transl. Genetics (Suppl) 35:30–32]
Wahlund S (1928) Zusammensetzung von Populationen und Korrelationsenscheinungen vom Standpunkt der Vererbungslehre aus betrachtet. Hereditas 11:65–106
Wallace ME (1965) Pseudoallelism at the agouti locus in the mouse. J Hered 56:267–271
Weinberg W (1908) Über den Nachweis der Vererbung beim Menschen. Jahresber Ver Vaterl Naturkd Württemb 64:368–382
Weinberg W (1909) Über Vererbungsgesetze beim Menschen. Z Indukt Abstamm Vererbungsl 1:227–330
Weinberg W (1910) Further contributions to the theory of inheritance. Arch Rassen Ges Biol 7:35–49
Wentworth EN, Remick BL (1916) Some breeding properties of the generalized Mendelian population. Genetics 1:608–616
Wildman J (1984) Major gene inheritance of head spots in mice. Thes Purdue Univ, Lafayette, Ind
Wildman J, Doolittle DP (1986) Single gene inheritance of head spot occurrence in mice. J.Hered 77:136–138
Wilson SP, Goodale HD, Kyle WH, Godfrey EF (1971) Long term selection for body weight in mice. J Hered 62:228–234
Wilson SP, Doolittle DP, Dunn TG, Malven PV (1972) Effects of temperature stress on growth, reproduction and adrenocortical function of mice. J Hered 63:324–330
Winchester AM (1951) Genetics. Houghton–Mifflin, Boston
Wright S (1920) The relative importance of heredity and environment in determining the piebald pattern of guinea pigs. Proc Natl Acad Sci USA 6:320–332
Wright S (1921) Systems of mating. Genetics 6:111–178
Wright S (1922) Coefficients of inbreeding and relationship. Am Nat 56:330–338
Wright S (1931) Evolution in Mendelian populations. Genetics 16:97–159
Wright S (1934) The method of path coefficients. Ann Math Stat 5:161–215
Wright S (1937) The distribution of gene frequencies in populations. Proc Natl Acad Sci USA 23:307–320
Wright S (1938) Size of populations and breeding structure in relation to evolution. Science 87:430–431
Wright S (1952) The theoretical variance within and among subdivisions of a population that is in a steady state. Genetics 37:312–321
Wright S (1955) Classification of the factors of evolution. Cold Spring Harbor Symp Quant Biol 20:16–24
Yule GU (1902) Mendel's laws and their probable relation to intra-racial heredity. New Phytol 1:193–207, 232–238

Subject Index

Population Genetics: Basic Principles

In contrast to earlier books on the subject, this compact text focuses on the potential for genetic manipulation. It provides the foundation in population and quantitative genetics, required to apply genetic principles to animal and plant breeding. Both lecturers and students will appreciate the convenient organization of the material, which makes this an excellent teaching text for a course in applied population genetics. With its concise treatment of basic concepts it should also provide a handy reference to established researchers.

ISBN 3-540-17326-9
ISBN 0-387-17326-9